4th International Conference on Upstream Engineering and Flow Assurance 2016

AA002655

Topical Conference at the 2016 AIChE Spring Meeting and 12th Global Congress on Process Safety

Houston, Texas, USA
10-14 April 2016

ISBN: 978-1-5108-2489-8

Printed from e-media with permission by:

Curran Associates, Inc.
57 Morehouse Lane
Red Hook, NY 12571

Some format issues inherent in the e-media version may also appear in this print version.

Copyright© (2016) by AIChE
All rights reserved.

Printed by Curran Associates, Inc. (2016)

For permission requests, please contact AIChE
at the address below.

AIChE
120 Wall Street, FL 23
New York, NY 10005-4020

Phone: (800) 242-4363
Fax: (203) 775-5177

www.aiche.org

Additional copies of this publication are available from:

Curran Associates, Inc.
57 Morehouse Lane
Red Hook, NY 12571 USA
Phone: 845-758-0400
Fax: 845-758-2633
Email: curran@proceedings.com
Web: www.proceedings.com

TABLE OF CONTENTS

An Innovative Advanced Control Solution for Slug Mitigation

Sanjay Sharma
Ravi Nath
Honeywell Process Solutions
Houston, TX

2016 AIChE Spring Meeting
Houston, TX
April 11, 2016

Honeywell

Agenda

- Introduction
- Traditional Solutions
  - PID
  - Choke Controller
- Proposed Solution
  - Slug Mitigation Controller
  - Case study
- Solution Performance
  - Test results
  - First project - progress report
- Conclusions

Introduction: Production Platform

2

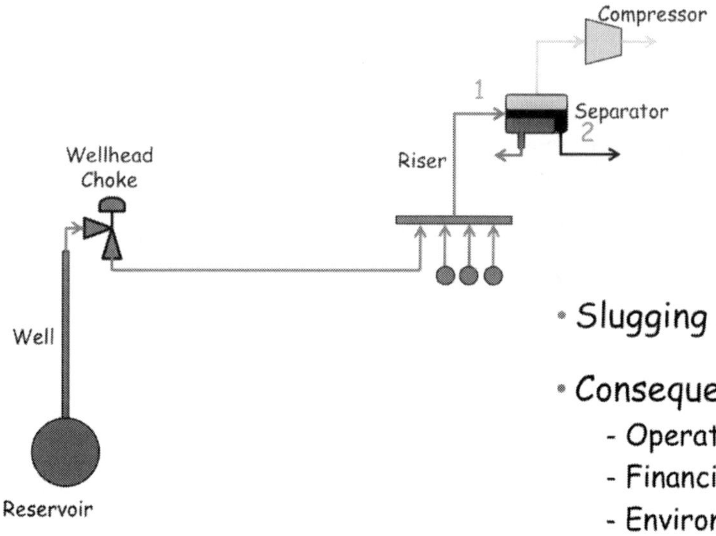

- Slugging happens
- Consequences
  - Operational
  - Financial
  - Environmental

Introduction: problem definition

3

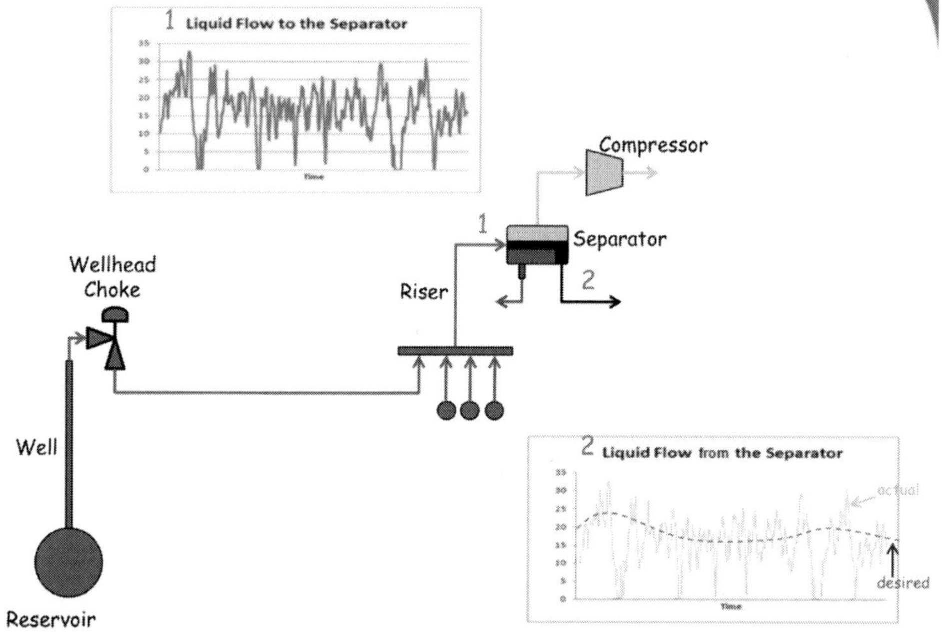

Introduction: Production Separator

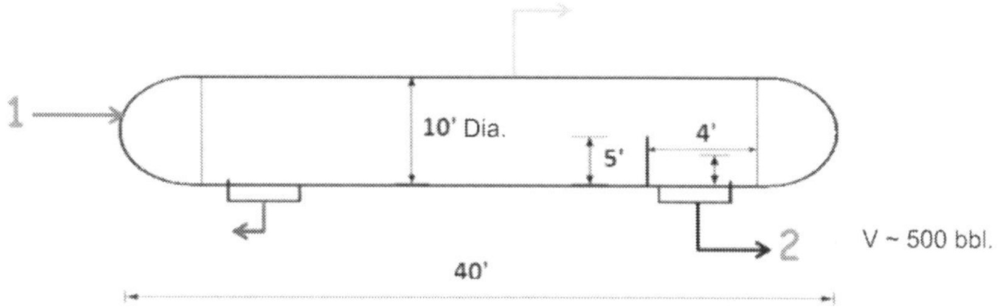

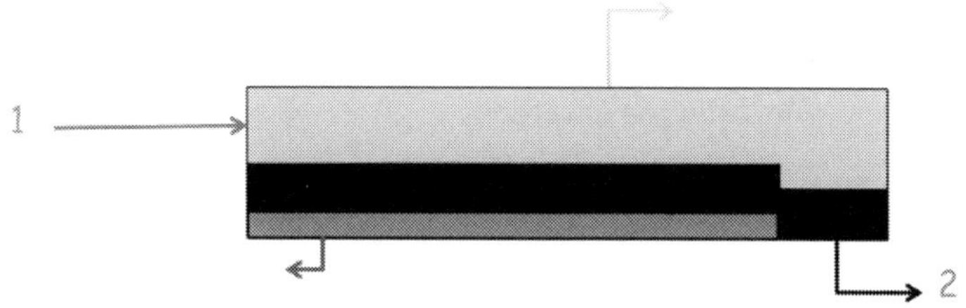

Traditional Solution: PID controller

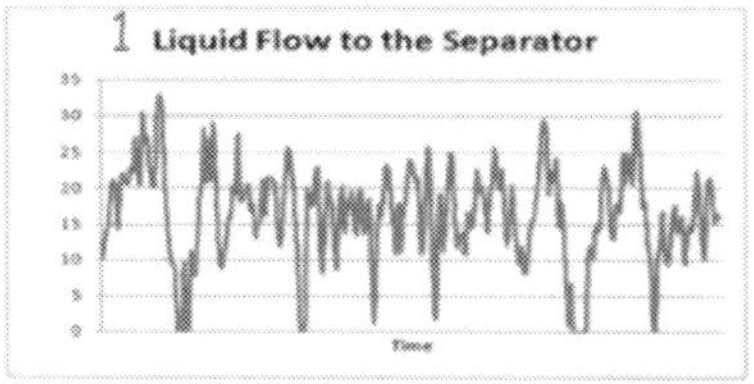

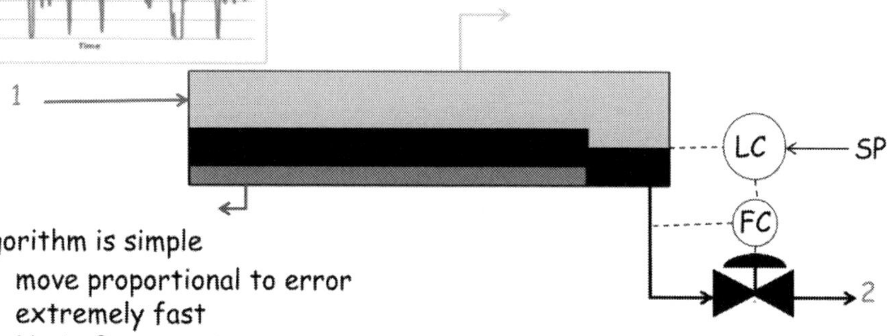

- Algorithm is simple
  - move proportional to error
  - extremely fast
  - Up to 3 parameters
  - Has no knowledge of limits
- Typically LC tuned very tightly
  - to prevent liquid carry-over

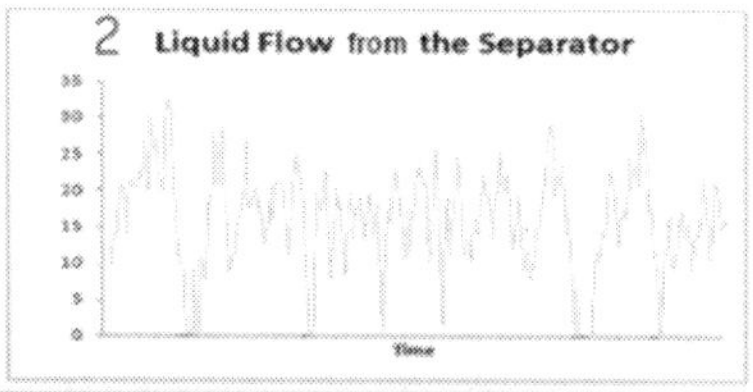

Minimal slug mitigation

Traditional Solution: Choking Controller

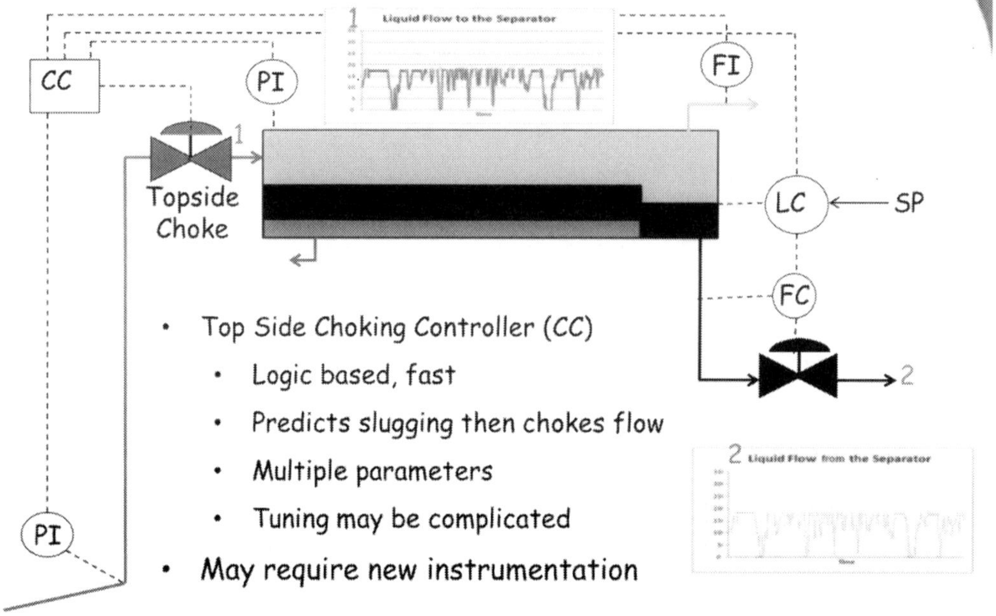

- Top Side Choking Controller (CC)
  - Logic based, fast
  - Predicts slugging then chokes flow
  - Multiple parameters
  - Tuning may be complicated
- May require new instrumentation

Mitigates slugging, but reduces production

Slug Mitigation: Proposed solution

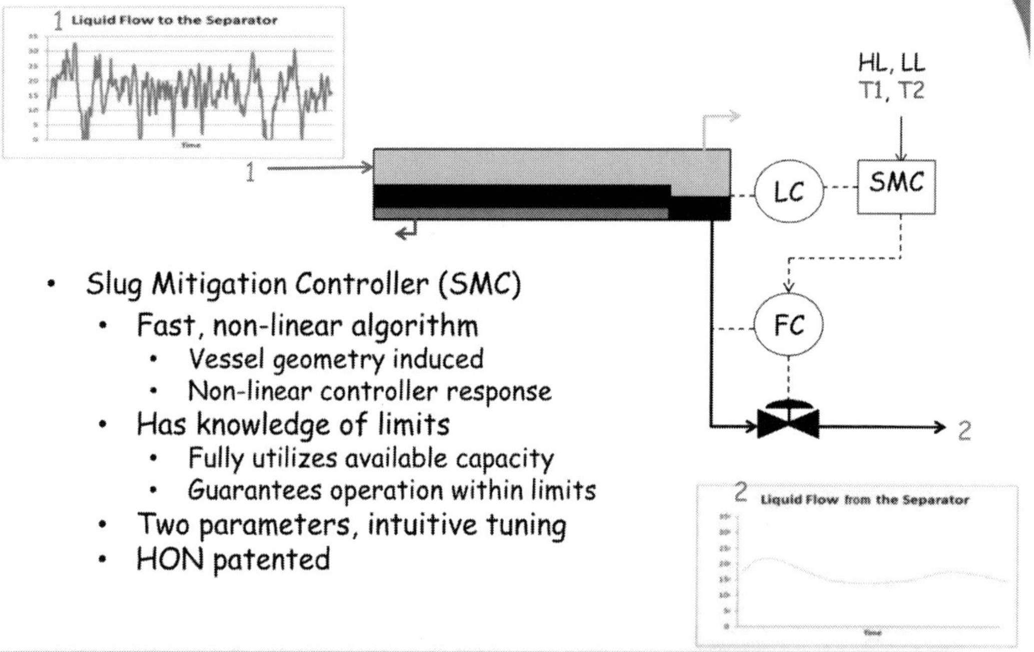

- Slug Mitigation Controller (SMC)
  - Fast, non-linear algorithm
    - Vessel geometry induced
    - Non-linear controller response
  - Has knowledge of limits
    - Fully utilizes available capacity
    - Guarantees operation within limits
  - Two parameters, intuitive tuning
  - HON patented

Does not choke production, generally no new instrumentation required

SMC algorithm (conceptual)

- Define
  - Capacity Deviation (CD): delta between actual inventory and limiting inventory
  - Setpoint Deviation (SD): delta between actual inventory and desired inventory

SMC algorithm (conceptual)

- Compute T* (unforced time to exhaust CD)
- If T* <= T1
  - Compute minimum move to exhaust CD in T1
- If T* > T1
  - Compute minimum move to exhaust SD in T2

SMC performance

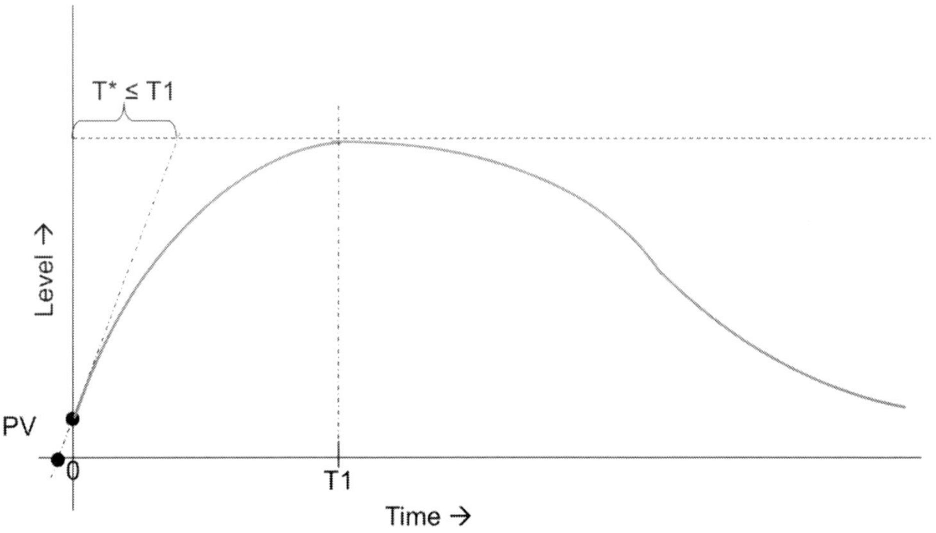

CD move calculation

SMC performance

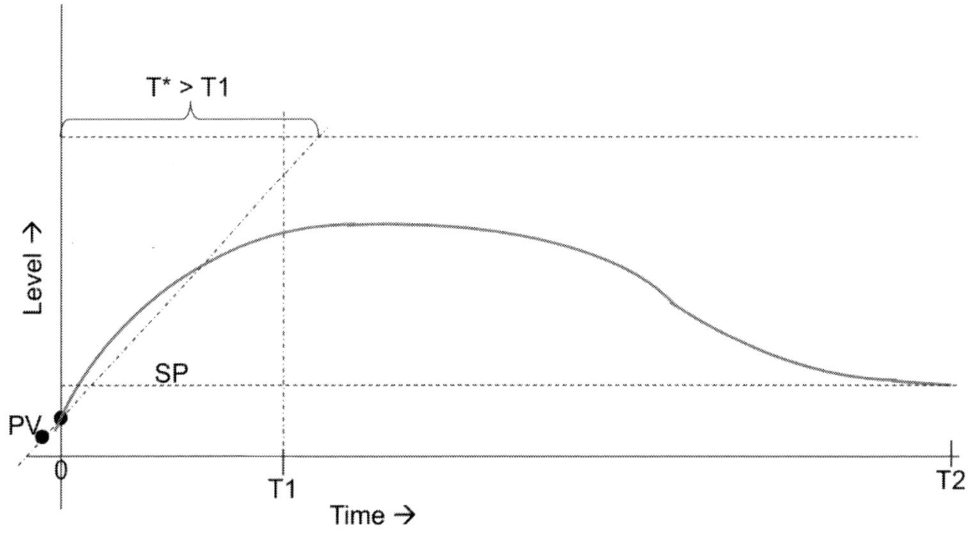

SD move calculation

SMC Test Results : GOM data

2 cascaded separators, flow variance reduction by 80%.

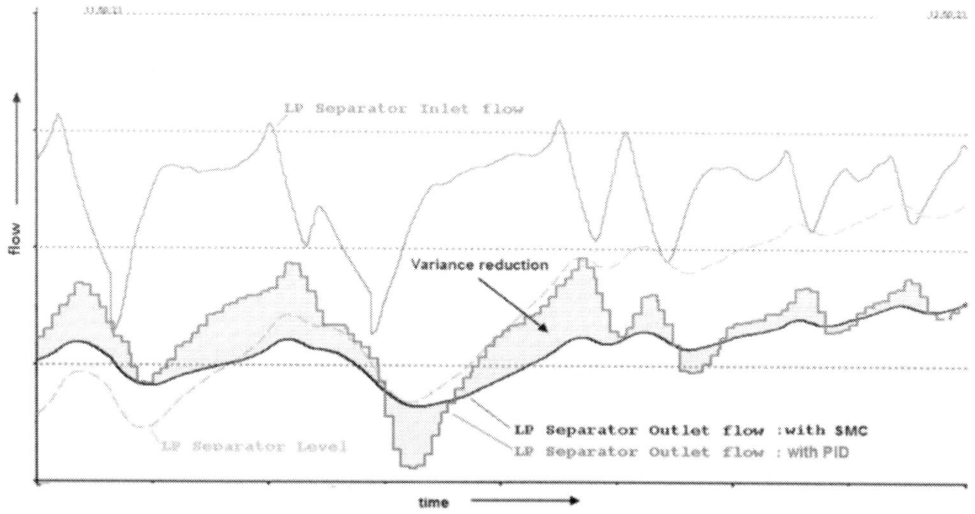

SMC First Project

Background:

- Overseas off-shore producer
- Old wells, very high water cut
- Hydro-cyclone is limiting production
- Would like to stabilize the flow to hydro-cyclone

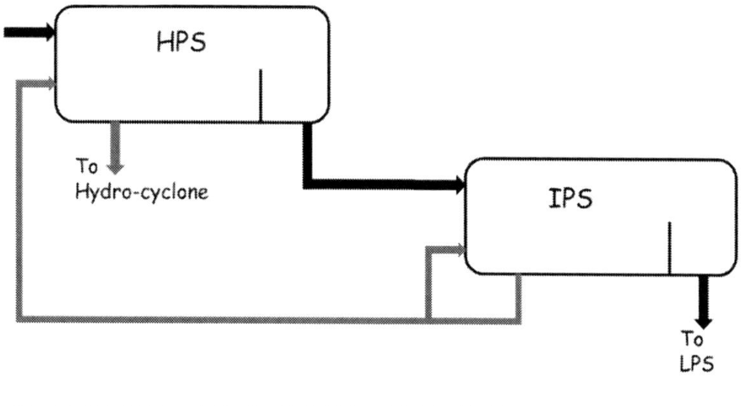

SMC First Project

Challenges:
- Water level measurement is inherently noisy
- Could SMC be applied to stabilize flow to hydro-cyclone?

Project status:
- Kick off:
  - just completed
  - Actual operating data has been collected
- Phase I
  - On going
  - Analyze data
  - Assess the suitability of SMC
  - Preliminary SMC design
  - Review scheduled for end of April '16
- Phase II
  - Implementation

Conclusions

- Slugging is disruptive to production
- Traditional solutions
  - PID
    - Little slug mitigation
  - Top side Choke Control (CC)
    - May require new instrumentation
    - Chokes production
- Proposed solution (SMC)
  - Patented, non-linear algorithm
  - Utilizes available separator capacity
  - Stabilizes outflow : significant variance reduction (up to 80%)
  - Generally no new instrumentation required
  - Does not choke production (no production reduction)

Q & A

16

444398 Hydrate Blockage Formation in Gas-Dominant Systems

Abstract:

Gas hydrates are ice-like solids, where molecular cages of water surround light hydrocarbon species at high pressure and low temperature. In subsea oil and gas flowlines, hydrate growth may result in constriction of the flow channel, increasing the frictional pressure drop across the line. In severe cases, hydrate formation may result in a complete blockage to production, requiring expensive remediation. Much of the benchtop and flowloop work to date has focused on hydrate formation from emulsified water in the oil phase, but there is insufficient data to inform hydrate growth rate or blockage mechanisms in gas-continuous systems. In this presentation, we propose a new conceptual model for hydrate blockage formation in gas systems, informed by the largest gas-dominant flowloop dataset collected to date. The mechanism includes a novel contribution for hydrate growth from liquid water dispersed in the gas phase, which provides a basis for sloughing behaviour observed in end-stage hydrate deposits. The growth model was benchmarked using a transient multiphase flow simulator (OLGA), and was able to accurately predict the average increase in frictional pressure drop within 0.2 psi/ft for high-velocity experiments. The effects of partial deposition and sloughing, hydrate film porosity, heat transfer resistance and surface roughness were not captured in the model, which represent a critical path forward for future experiments.

Copyright © American Institute of Chemical Engineers. All rights reserved.

16AIChE
Spring Meeting, Houston, TX

Pressure and Temperature Optimization for Fluids Produced by Deepwater Oil or Gas Wells: a Study on Current Achievements and Possible Future Developments

Francesco Beltrami
Granherne Inc., a KBR Company
601 Jefferson, Houston, Texas
francesco.beltrami@kbr.com

© 2016, KBR Inc. All Rights Reserved

Prepared for Presentation at
American Institute of Chemical Engineers
2016 Spring Meeting
11th Global Congress on Process Safety
Houston, Texas
April 10-14, 2016

UNPUBLISHED

AIChE shall not be responsible for statements or opinions contained in papers or printed in its publications

Pressure and Temperature Optimization for Fluids Produced by Deepwater Oil or Gas Wells: a Study on Current Achievements and Possible Future Developments

Francesco Beltrami
Granherne Inc., a KBR Company
601 Jefferson, Houston, Texas
francesco.beltrami@kbr.com

Keywords: Deepwater Wells, Subsea Production System, Multiphase Boosting, Subsea Heating

Abstract

To overcome the technical challenges associated to deepwater wells production, and to ensure a greater recovery of hydrocarbons during the field life, actions can be taken to modify composition, flow rate, pressure and temperature of the produced fluids inside the subsea production system, from the wells up to the host.
This paper reviews the current ways of increasing or decreasing in a controlled way pressure or temperature of the produced fluids and discusses possible technical solutions that may have to be developed in the future. Ways to improve the transportation of hydrocarbons over long distances are investigated and the impact of new technologies on the current way of producing deepwater oil and gas wells is assessed.

1. Introduction

1.1 Overview

The growth of the subsea industry over the past thirty (30) years has been remarkably successful, enabling the development of several offshore oil and gas fields in increasingly deeper waters, harsher environments and more challenging conditions. Oil and gas produced from deepwater fields will continue in the future to be a significant component of the overall global energy supply. Past and current technical achievements of key subsea technologies and systems have to be taken into account when looking at the different development options for a given offshore field: only components that offer the required technical readiness level and systems that have undergone complete qualification programs can be specified and used.
Notwithstanding the large amount of information nowadays available in the public domain on projects, it is not always clear to which extent a technology is properly applicable, for example, the fact that one component has been installed in shallow water does not mean it can work in deeper waters or with different produced fluids.

1.2 Study Objectives

The aim of this paper is to provide precise indications on some selected technologies that have been, are, or might be used to design subsea production systems (SPS's) for deepwater and ultra-deepwater oil and gas fields. Practical ways of decreasing or increasing in a controlled way the pressure or the temperature of the produced fluids are reviewed. Future applications are also discussed to visualize expected changes to technology boundaries and possible evolution lines. The main subsea technologies discussed in the paper are:

- Pressure reduction: choke valve, pipeline protection system, high integrity pressure protection system (HIPPS), flow restrictor (future)
- Temperature reduction: cooler, bundle cooling, spool cooling
- Pressure increase: multiphase boosting, wet gas compression
- Temperature preservation or increase: flowline and riser heating, thermal insulation, local heating (future).

1.3 System Description

The paper focuses on the typical deepwater SPS designed in accordance with [1] or [2] to handle reasonably 'common' fluids in terms of physical properties (max 15,000 psi pressure rating; 121°C temperature) and chemical components - equipment designed for sour service as per [3]. However, some high pressure high temperature (HPHT) projects are mentioned here because they are likely to become more frequent in the future.

1.4 Exclusions

Only information available in the public domain has been used to write this paper: operators and manufacturers have not been contacted at this time. The status of the technologies under investigation [4] and of relevant field development options [5] is available in the public domain and will not be presented here. Recent information on subsea processing projects and their corresponding hosts is available in [6].
There are several parameters that need to be examined when assessing the technology readiness status of a given SPS components, but only the water depth (WD) is discussed here. Water depth is a key indicator of technical features, design requirements and practical performances, and it applies to all the technologies discussed here. The limit between shallow water and deepwater is here assumed equal to 300m. The transition between deepwater and ultra-deepwater takes place at 1,500m. These boundaries differ from those indicated in [7], but are used in the industry and better reflect the fact that that divers cannot work in deepwater.

2. Current Achievements

2.1 Pressure Decrease

The first subsea choke valve was installed in 1975 in the Gulf of Mexico (GoM), see [8]. Choke valves are typically part of the subsea tree system for production, water injection or gas injection, either placed on the tree itself or located in a flow module that can be retrieved to surface. It is understood that the deepest subsea tree currently in operation is equipped with a subsea choke valve [9]. Choke valves may be also mounted on subsea manifolds, see for example [10].
The first HIPPS in operation is conventionally considered to be Kingfisher in the North Sea: the field start-up was in 1997 [11]. The first 'pipeline protection system' in the GoM is Mensa that started operations in 1998. Mensa is a record-breaking subsea tieback of deepwater gas wells to an existing shallow water host [12]. The deepest HIPPS is possibly the one installed at Mexilhão in the Brazilian offshore at 500-600m water depth [13]. The Mexilhão HIPPS is located on the subsea tree. Another deepwater 'pipeline protection system' is the gas export regulating manifold (GERM) in the Block 18 Gas Export Line to Angola LNG [14].

The only example of a flow restrictor encountered in this study is the component belonging to the Marlim subsea separation system [15] designed to handle heavy oil and installed in 870m WD in the Brazilian offshore.

2.2 Temperature Decrease

The first known applications of subsea heat exchangers are Cavendish, in the North Sea and Kipper in Southern Australia. Cavendish water depth is 18.5m. Gas production started in 2007. Kipper is also a gas field, but in deeper waters (100m WD), see [16]. It started production in 2013. In these and in all the other cases reviewed the cooling medium is seawater. Subsea cooling systems have been adopted in the design of bundles: the first was possibly Jura in the North Sea [17], in 113m WD. The subsea cooling system is a part of a bundle towhead. Production started in 2008. The deepest of such systems appears to be in Knarr, still in the North Sea, in 410m WD (start-up in 2015).
Subsea coolers are also installed in the Asgard [18] and Gullfaks [19] wet gas compression systems that started operations in late 2015. Both fields are in the North Sea. Asgard's water depth is 270m while Gullfaks' water depth is 125m. Subsea wet gas compression was also qualified for Ormen Lange in the North Sea (850m WD). A pilot was extensively tested onshore till 2014 [20] but its full commercial application has not been sanctioned [21].
The use of cooling spools, such as an uninsulated section of a production flowline, has been investigated in the past and in more recent times [22], but practical oilfield use has not been encountered.

2.3 Pressure Increase

The first subsea commercial application of a multiphase booster on the seabed started its operations in Draugen, North Sea, in 1994: it was a turbine-driven machine installed in 270m WD [23]. The first multiphase boosters coupled to electric motors were used in Lufeng, South China Sea, at 330m WD, between 1997 and 2009 [24]. Still in 1994 the first electrical submersible pump (ESP) in a subsea well started its operations at Carapeba, 90m WD, in the Brazilian offshore [25].
The deepest subsea multiphase booster currently installed is the Jack St. Malo (JSM) pump in 2,134m WD in the GoM, see [23]. The field started operations in 2014. The JSM pump is rated 13,500 psi; this is currently the highest pressure rating in use. Three (3) ultra-deepwater projects have adopted ESP's, although in different configurations:

- ESP on the seabed, Cascade & Chinook, GoM, in approx. 2,900m WD, see [26]
- ESP within subsea boosting station, Perdido, GoM, 2,383m WD, [27]
- ESP in a caisson, BC-10, Brazilian offshore, 1,900m WD max [28].

The first subsea wells completed with hydraulic submersible pumps (HSP) were constructed in Captain, North Sea, in approx. 120m WD. The field started production in 1997 [29]. The second known application of HSP's in subsea wells is Kraken, North Sea, 125m WD, currently under development: start up is expected in 2016-17 [30].
The use of jet pumps/ ejectors to boost the fluids produced by subsea wells has been investigated, but its practical use has been limited so far. Jet pumps are part of the heavy oil subsea separation system installed in Marlim, offshore Brazil, in 870m WD [15]. Another known subsea-related application of jet pumps is Gemini. In this case, the jet pumps were installed on the host topsides and proved to be beneficial in increasing the productivity of the subsea wells located at 1,115m

WD [31]. The only two (2) existing commercial applications of subsea wet gas compressors were mentioned in section 2.1 above. It is premature to discuss how they are performing in practice.

2.4 Temperature Preservation or Increase

Different solutions are possible to prevent the unwanted cooling of the fluids transported inside the flowlines, see [32]. In the North Sea, bundles have been adopted for this purpose. One of the advantages offered by bundles is that they make possible the use of thermal insulation that cannot be exposed to seawater [33]. Bundles can also be heated: the first bundle of this type is probably the one installed in Britannia in 150m WD: the heating medium was hot water pumped from the host topsides [34].

Pipe-in-Pipe flowlines can be designed to be heated with a cooling medium, such as in King, GoM, in 1,500m WD: in this case, the medium is aqueous ethylene glycol solution with inhibitors [35]. The field started its operations in 2012.

The first flowline equipped with Direct Electrical Heating (DEH) was installed in Asgard, in the North Sea, in 270m WD. Production started in 2000 [36]. The deepest monobore flowline equipped with a DEH system has been recently installed in Lianzi, Cabinda (West Africa) in 1,070m WD [37]. The deepest Pipe-in-Pipe flowline with a DEH system was installed in 2004 in Na Kika, GoM, in 1,800m WD [38].

Electrical Trace Heating (ETH) is getting wider acceptance too: the first ETH flexible pipe is in use in Dalia in the Angolan offshore in 1,350m since 2006 [39]. ETH flexible pipe has been used in other projects too, but Dalia remains the deepest at this time of writing. The first ETH Pipe-in-Pipe was used for Islay, North Sea, in 120m WD. It started operations in 2012 [39].

To date, the deepest Pipe-in-Pipe flowline system is the one installed in 2012 in Cascade & Chinook, GoM, in 2,900m WD [40]. The inner pipe is thermally insulated by means of ultra-low-conductivity materials, while the outer pipe provides mechanical protection.

3. Future Developments

3.1 Pressure Decrease

It is likely that in the future subsea choke valves will be used in deeper waters. One possible application might be Stones, GoM, in approx. 2,900m WD. More difficult is to foresee when subsea choke valves with a 20,000 psi pressure rating will be available on the market. In case of very high pressures, it might become necessary to use two choke valves in series [41].

HIPPS is being used in deeper waters and in different geographic areas. A deepwater application will be in Julia, GoM, in 2,460m WD [42]. A possible technological development is the 'all-electric' HIPPS, see [43].

3.2 Temperature Decrease

Developing future HPHT fields will require the adoption of subsea coolers, as stand-alone units on the seabed (for example, Alder and Linnorm in the North Sea), or as cooling spools integrated in a bundle towhead. Subsea coolers will be used also in future subsea wet gas compression systems, possibly in deeper waters (Ormen Lange in 850m WD?)

3.3 Pressure Increase

According to press releases and other news available in the public domain, the first subsea multiphase booster rated 15,000 psi will be part of Stones, GOM. The water depth of this project is 2,900m. The field start-up is planned for 2016.

ESP are likely to be used in future deepwater field development, not only in the arrangements already experimented, see section 2.3, but also on rigid jumpers [44] or in production risers [45]. Very limited information has been encountered on possible future applications of HSP's, with the exception of Kraken (see section 2.3) and jet pumps. More is available in the public domain on upcoming subsea wet gas compression projects in deeper waters, see [4]. Late life compression for gas field remains a challenging issue (subsea versus topsides), but due to the current market conditions, it might take time before operators sanction new subsea wet gas compression projects, especially in deepwater environments.

3.4 Temperature Preservation or Increase

A future DEH system might be deployed in 2,300m WD offshore Brazil [46]. A deeper ETH Pipe-in-Pipe flowline is understood to be under development for a project in West Africa [47]. In this case the water depth should be approximately 700m. It can be also expected that more projects will adopt ETH flexible pipe. It will be interesting to learn about the Stones flowlines and risers, when information will be publicly disclosed.

Heating the well jumper has been proposed [48] but no information was found on possible subsea heaters. There might be different reasons to explain why the industry is not looking at a subsea heater; its complexity and cost, the need for a heating medium to be supplied, the risk of fouling and other maintenance aspects, the fact that heating the flowlines may be a satisfactory solution.

In the early stages of the subsea industry, nuclear power was used to operate subsea trees and other subsea equipment, see [49] and [50]. The industry paid great attention to the possible ways of generating power on the seabed: alternatives were explored, see for example [51], [52] and [53], and some were tested in practice [54]. The solution that emerged, however, and become standard, was to generate power on the host, or on the coast, and then distribute the power at the field locations on the seabed.

Looking at the increasing complexity of subsea power distribution over long distances, the greater power required, and the availability of new power sources (renewables and others) it might be worth exploring if power, or even heat, can be simply and reliably generated on the seabed to heat the produced fluids, preventing hydrates formation or solids deposition. In doing so, looking at other industries, such as aerospace, defense and renewable energy may be beneficial.

4. Conclusions

Selected technologies that are employed in current SPS's design were reviewed. Future possible applications were also investigated to determine how the applicability of such technologies may evolve over the time. The main findings of this study are:

- Pressure reduction: technology is mature and may move into deeper waters in the future
- Temperature reduction: the industry is gaining experience with controlled cooling of the produced fluids. The technology is field proven is shallow water. It has to be seen if and when it will move to deeper waters
- Pressure increase: subsea multiphase boosting continues its successful history. Subsea wet gas compression is relatively new and needs to be proven in practice before moving into deepwater
- Temperature preservation or increase: well-established heating techniques will continue to be employed. It can be expected that both heating techniques and thermal insulation will continue their development and will be adopted in even deeper waters.

Subsea heaters appear to not be in use. Taking into account the progress made by the subsea technologies and by other industries in the past decades, it might be now possible to produce heat and use it in a controlled way to increase the temperature of the produced fluids. The challenge here is to identify a simple, reliable and affordable system to validly compete against the potential use of high voltage and high power subsea electric grids.

5. References

[1] API 17, "Design and Operation of Subsea Production Systems" (entire suit of documents)

[2] ISO 13628, "Petroleum and Natural Gas Industries - Design and Operation of Subsea Production Systems" (entire suit of documents)

[3] ISO 15156, "Petroleum and Natural Gas Industries - Materials for Use in H_2S-Containing Environments in Oil and Gas Production" (comes in three parts)

[4] "2016 Worldwide Survey of Subsea Processing: Separation, Compression, and Pumping Systems", Offshore Magazine, March 2016

[5] "2015 Deepwater Solutions and Records for Concept Selection", Offshore Magazine, May 2015

[6] Anikpo, A., and Beltrami, F., "Subsea Processing versus Host Selection: an Imperative Correlation". Offshore Technology Conference, 04-07 May 2015, Houston, Texas, USA. Paper 25675-MS

[7] API RP 17A, "Design and Operation of Subsea Production Systems - General Requirements and Recommendations"

[8] "Our history", https://www.c-a-m.com/company (accessed Feb. 22, 2016)

[9] "Perdido World's Deepest Offshore Drilling and Production Platform", supplement to Offshore Magazine, April 2010

[10] Couto, P., Moreira, C., Ribeiro, L., Dantas, S., and Waskoviak, R., "Albacora Manifold - The Deepest Subsea Manifold Installed To Date". Offshore Technology Conference, 6-9 May 1996, Houston, Texas, USA. Paper OTC-8236-MS

[11] Hutchings, V., "Is the Subsea High Integrity Pressure Protection System (HIPPS) Coming of Age?" Society of Underwater Technology, Subsea Control and Data Acquisition (SCADA) Conference, 2-3 June 2010, Newcastle, UK. Paper SUT-SCADA-10-169

[12] Gilchrist, R. T., and Kluwen, F. A., "Mensa Project: Flowlines". Offshore Technology Conference, 4-7 May 1998, Houston, Texas, USA. Paper OTC-8628-MS

[13] "Petrobras Mexilhão", http://www.fmctechnologies.com/~/media/Subsea/Projects/Brazil/Petrobras%20Mexilhao_new.ashx?force=1&track=1 (accessed Feb. 22, 2016)

[14] Oldfield, G., "Subsea, Umbilicals, Risers and Flowlines (SURF): Performance Management of Large Contracts in an Overheated Market; Risk Management and Learnings". Offshore Technology Conference, 5-8 May 2008, Houston, Texas, USA. Paper OTC-19676-MS

[15] Orlowski, R., Euphemio, M. L. L., Euphemio, M. L., Andrade, C. A., Guedes, F., Tosta da Silva, L. C., Pestana, R. G., De Cerqueira, G., Lourenço, I., Pivari, A., Witka, A., Folhadella, H., Pacheco, L., Kronemberger, S., and Vilela, J., "Marlim 3 Phase Subsea

Separation System - Challenges and Solutions for the Subsea Separation Station to Cope with Process Requirements". Offshore Technology Conference, 30 April-3 May 2012, Houston, Texas, USA. Paper OTC-23552-MS

[16] Douglas, B., Pugh, D., Monahan, C. C., and Russ, P. R., "Top of Line Pipeline Corrosion Management through Innovate Subsea Facility Design". International Petroleum Technology Conference, 3-5 December 2008, Kuala Lumpur, Malaysia. Paper IPTC-12788-MS

[18] Hedne, P., "Managing the Risk of the Unknowns: Åsgard Subsea Compression Qualification Program". Offshore Technology Conference, 05-08 May 2014, Houston, Texas, USA. Paper OTC-25409-MS

[19] Hjelmeland, M., Olsen, A. B., and Marjohan, R., "Advances in Subsea Wet Gas Compression Technologies". International Petroleum Technology Conference, 15-17 November 2011, Bangkok, Thailand. Paper IPTC-14231-MS

[20] Skofteland, H., Hilditch, M., Normann, T., Eriksson, K. G., Nyborg, K., Postic, M., and Camatti, M., "Ormen Lange Subsea Compression Pilot - Subsea Compression Station". Offshore Technology Conference, 4-7 May 2009, Houston, Texas, USA. Paper OTC-20030-MS

[21] Rijnbeek, R., "Sub Sea Compression at Ormen Lange". International Petroleum Technology Conference, 19-22 January 2014, Doha, Qatar. Paper IPTC-17649-MS

[22] Animah, I., and Addy-Lamptey, A., "Design Considerations of Subsea Cooling Spool", Innovative Systems Design and Engineering, Vol. 6, No. 1, 2015

[23] Solvik, N. A., Michaelsen, J., and Luce, P., "Evolution of Subsea Boosting: A Review of Past and Present Projects". Offshore Technology Conference, 6-9 May 2013, Houston, Texas, USA. Paper OTC-24263-MS

[24] Davies, S. R. H., Bakke, W., Ramberg, R. M., and Jensen, R. O., "Experience to date and future opportunities for subsea processing in StatoilHydro". Offshore Technology Conference, 3-6 May 2010, Houston, Texas, USA. Paper OTC-20619-MS

[25] Mendonca, J. E., Hedge, R. C., Izetti, R., Nicholson, A., Morrison, D., Dwiggins, J. L., Cia, M., and Alfano, P. P., "First Installation of an Electrical Submersible Pump in a Subsea Well". Offshore Technology Conference, 1-4 May 1995, Houston, Texas, USA. Paper OTC-7907-MS

[26] Cassity, T. G., O'Donnell, M., Weigle, J., Masson, C., Kochi, K., and Porciuncula, S. L., "Cascade and Chinook Subsea Development: A Challenging and Successful Case History". Offshore Technology Conference, 6-9 May 2013, Houston, Texas, USA. Paper OTC-24162-MS

[27] Ju, G. T., Littell, H. S., Cook, T. B., Dupre, M. H., Clausing, K. M., Shumilak, E. E., Schoppa, W., and Blizzard, W. A., "Perdido Development: Subsea and Flowline Systems". Offshore Technology Conference, 3-6 May 2010, Houston, Texas, USA. Paper OTC-20882-MS

[28] Sleight, N. C., and Oliveira, N., "BC-10 Optimizing Subsea Production". Offshore Technology Conference Brazil, 27-29 October 2015, Rio de Janeiro, Brazil. Paper OTC-26220-MS

[29] Mali, G. A., Morrison, A. K., Green, A., Graham, R., and Harden, W. G., "Hydraulic Submersible Pumps: 10 Years Experience on a Heavy-Oil Field in the North Sea". SPE

Annual Technical Conference and Exhibition, 19-22 September 2010, Florence, Italy. Paper SPE-135234-MS

[30] Downie, A., and Bhatia, A., "The HSP Versatile Artificial Lift Choice for the Heavy Oil Subsea Kraken Development", https://www.spe-uk.org/aberdeen/wp-content/uploads/2014/07/16.-The-HSP-Versatile-Artificial-Lift-Choice-for-the-Heavy-Oil-Subsea-Kraken-Development-EnQuest.pdf (accessed Mar. 3, 2016)

[31] Becnel, M., Steib, D., Wegener, S., Linder, J., and Beg, N., "Maintaining Late-Life Production From Long Offset Subsea Gas Wells". Offshore Technology Conference, 30 April-3 May 2007, Houston, Texas, USA. Paper OTC-18676-MS

[32] Ansart, B., Marret, A., Parenteau, T., and Rageot, O., "Technical and Economical Comparison of Subsea Active Heating Technologies". Offshore Technology Conference Asia, 25-28 March 2014, Kuala Lumpur, Malaysia. Paper OTC-24711-MS

[33] Goodlad, M., "Bundle Technology for the Future". Society of Petroleum Engineers Offshore Europe Oil and Gas Conference and Exhibition, 3-6 September 2013, Aberdeen, UK. Paper SPE-166561-MS

[34] Brown, L. D., Clapham, J., Belmear, C., Harris, R., Loudon, A., Maxwell, S., and Stott, J., "Design of Britannia's Subsea Heated Bundle for a 25 Year Service Life". Offshore Technology Conference, 3-6 May 1999, Houston, Texas, USA. Paper OTC-11017-MS

[35] Esaklul, K. A., Fung, G., Harrison, G., and Perego, R., "Active Heating for Flow Assurance Control in Deepwater Flowlines". Offshore Technology Conference, 5-8 May 2003, Houston, Texas, USA. Paper OTC-15188-MS

[36] Dretvik, S., and Bornes, A. H., "Direct Heated Flowlines in the Asgard Field". International Society of Offshore and Polar Engineers, 11th International Offshore and Polar Engineering Conference, 17-22 June 2001, Stavanger, Norway. Paper ISOPE-I-01-235

[37] "Chevron's Lianzi Sets New DEH Record", http://www.epmag.com/chevrons-lianzi-sets-new-deh-record-826616 (accessed Mar. 12, 2016)

[38] Kopp, F., Light, B. D., Preli, T. A., Rao, V. S., and Stingl, K. H., "Design and Installation of the Na Kika Export Pipelines, Flowlines and Risers". Offshore Technology Conference, 3-6 May 2004, Houston, Texas, USA. Paper OTC-16703-MS

[38] Gloaguen, M., Bourdillon, H., Roche, F., de Reals, T. B., Menier, P., and Marion, A., "Dalia Flowlines, Risers, and Umbilicals". Offshore Technology Conference, Offshore Technology Conference, 30 April-3 May 2007, Houston, Texas, USA. Paper OTC-18543-MS

[39] Nebell, F., De Naurois, H. J., Parenteau, T., and Decrin, M. K., "Flow assurance modeling using an Electrical Trace Heated Pipe-In-Pipe: From Qualification to Offshore Testing". Offshore Technology Conference, 6-9 May 2013, Houston, Texas, USA. Paper OTC-24060-MS

[40] Novelli, P., Armengol, M., Darcis, P., Anelli, E., Fazackerley, B., Kristiansen, P., and Blockhus, D., "Cascade and Chinook Project: Pipes and Bends: Material Design and Production". Offshore Technology Conference, 2-5 May 2011, Houston, Texas, USA. Paper OTC-21832-MS

[41] Meng, W., and Prescott, C. N., "Dual Subsea Choke Applications for High Pressure Reservoir Development". Society of Petroleum Engineers International Oil Conference and Exhibition in Mexico, 27-30 June 2007, Veracruz, Mexico. Paper SPE-108432-MS

[42] Dent, S., "Deepwater Gulf of Mexico - Julia Project HIPPS". Subsea Tieback Forum, 3-5 March 2015, New Orleans, Louisiana, USA

[43] Lamison, C. W., and Mateen, K., "All Electric Subsea HIPPS Technology Readiness". Deep Offshore Technology Conference, Perth, Australia, 27-29 November 2012. Paper T3S4O4

[44] Homstvedt, G., Pessoa, R., Portman, L., Wang, S., Gonzalez, J., Maldaner, M., and Margulis, J., "Step-Change Seabed ESP Boosting". Offshore Technology Conference Brazil, 27-29 October 2015, Rio de Janeiro, Brazil. Paper OTC-26141-MS

[45] Rodrigues, R., Foresti, B. P., Vieira, T. S., and de Franca, A. B., "A New Boosting Concept: Pumps Installed inside Production Risers in FPSOs". Offshore Technology Conference Brazil, 27-29 October 2015, Rio de Janeiro, Brazil. Paper OTC-26129-MS

[46] "Brazilian offshore oil field takes Nexans' DEH system to record depths" http://www.nexans.no/eservice/Norway-en/navigatepub_142640_-34367/Brazilian_offshore_oil_field_takes_Nexans_DEH_syst.html (accessed Mar. 8, 2016)

[47] McDermott, P., and Sathananthan, R., "Active Heating for Life of Field Flow Assurance". Offshore Technology Conference, 05-08 May 2014, Houston, Texas, USA. Paper OTC-25107-MS

[48] Bardon, F., Hudson, W. L., and Cretenet, A., "Hydrate Prevention with Electrically Heated Jumpers". Offshore Technology Conference, 30 April-3 May 2007, Houston, Texas, USA. Paper OTC-18773-MS

[49] Goodfellow, R., Webb, A., and Harbonn, J., "Subsea Experience Gained at Zakum". Offshore Technology Conference, 6-8 May 1974, Houston, Texas, USA. Paper OTC-1942-MS

[50] Mahler, C., and Awo, R., "The Future of Electric Controls: Trees and Subsea Processing". Offshore Technology Conference Asia, 25-28 March 2014, Kuala Lumpur, Malaysia. Paper OTC-25009-MS

[51] Young, H. L., "Development in Subsea Controls". Society of Petroleum Engineers, Offshore Europe, 5-8 September 1989, Aberdeen, United Kingdom. Paper SPE-19226-MS

[52] Byrne, S., Juguet, J.-P., and Dubourdieu, P., "APAC Concept: A New Umbilical-Less Subsea Control System". SUT Subsea Controls and Data Acquisition: Cost effective challenges for a geographically expanding industry, 22-23 April 1998, London, UK. Paper SUT-SCADA-98-119

[53] Maneschi, G., Menegazzo, F., Chimisso, C., and Mattei, L., "A Gas Turbine Power Generator for Improving Autonomous Subsea Control Systems". Offshore Technology Conference, 3-6 May, Houston, Texas, USA. Paper OTC-7256-MS

[54] Galletti, R., Citi, G., and Battaia, C., "Sea Water Batteries Application to the Luna 27 Autonomous Well". Society of Petroleum Engineers, European Petroleum Conference, 22-24 October 1996, Milan, Italy. Paper SPE-36940-MS

444408 Discussions on Design of Deepwater Flow Assurance of LW3-1 Gas Field

Abstract:

ABSTRACT.

With the increase of the water depth, the depth and length of subsea pipelines also increase. The hard circumstances of deepwater areas not only make harsh requirements for deepwater floating platforms, subsea production facilities and offshore operations, but also bring huge challenges for subsea pipelines from several kilometers to few hundreds kilometers, which tie subsea satellite wells, marginal fields to central processing systems. The complex of the gas oil and water multiphase flow, seabed hypsographic feature and operation activities etc., bring a series of problems to subsea transportation system, such as solid formation (hydrate, wax), severe slug flow in deep water riser, multiphase flow corrosions and solid particle's erosion etc. These problems have seriously affected general running and safety of the subsea production oil fields, especially in the deepwater multiphase transportation systems, and accidents induced by these problems happen repeatedly. This paper summary the typical flow assurance problems related to the Chinese first deepwater gas field LW3-1. On this base, the flow assurance problems about the 79 Km multiphase flow along the slope and its design methods are presented.

Copyright © American Institute of Chemical Engineers. All rights reserved.

445323 Water-Flexible Fracturing Systems

Abstract:

Developed to mitigate the economic burden of freshwater transportation and subsequent disposal, novel water- flexible fracturing systems use non-freshwater sources for hydraulic fracturing operations. This type of "fluid-first" solution can reduce or eliminate freshwater use and its associated transportation and disposal costs, while decreasing environmental impact. Water-flexible fluid systems have been tailored even to the most difficult water sources and require little or no treatment—and no disposal. Rather than focusing on expensive treatments to meet fracturing fluid specifications, the fluid-first holistic water management approach looks to fluid upgrades to replace treatment. The integrated approach requires detailed water characterization data, innovative fluid design, proactive scale modeling, analysis of rock-fluid interactions, and comprehensive assessment of full water-cycle costs. Benefits of the approach have included superior proppant placement, clay stability without additional stabilizers, no operational issues, and no negative impact on production rates in field trials. In addition, some surprising benefits were realized including less pumping horsepower required, higher fluid viscosity, and increased production. From a financial perspective, the overall water cycle costs were reduced, with fewer trucks on the road for transport, and disposal costs were eliminated. These engineered systems use extensive information gained from field, water, and reservoir characterization exercises, laboratory testing, and modeling to plan, execute, and evaluate the success of the overall water management program. By accommodating alternative sources of water, these systems can provide superior economics and allow operators to reuse produced water and other non-freshwater sources obtained near-site, while greatly reducing water handling, transportation, and disposal well costs. In addition, as stricter regulations for freshwater use and disposal go into effect, operators can prevent downtime by proactively lowering their environmental impact. Actual case studies utilizing water-flexible fracturing systems demonstrate the technical and economic benefits of this approach.

Copyright © American Institute of Chemical Engineers. All rights reserved.

445666 Optimizing Hydraulic Fracturing using Material Point Method Geomechanical Tool

Abstract:

Hydraulic fracturing is the process that allows economic development of unconventional reservoirs. The process is very challenging and complex and is characterized by fracture propagation. New fractures are formed because of the stress field resulting from the stimulation of a fractured rock formation in a complex geological setting with discontinuities and interfaces, varying material properties, etc. These reservoir features impact in specific ways the stress field prior to and during hydraulic fracture propagation, and thus impact the final network of hydraulic and natural fractures that determine the shale permeability and outcome. One way to optimize the hydraulic fracturing process is for completion operators to know the stress field prior to stimulation so they can plan the position of the completion stages in areas where the stimulation will be favorable, allowing for complex fracture propagation.

This presentation focuses on the evaluation of the stress anisotropy resulting from the presence of natural fractures and facies in a reservoir. The geomechanical tool, Material Point Method, was used as a means of assessing the stress state in a fractured and heterogeneous media. The natural fracture network and the facies were evaluated using geostatistical algorithms. The resulting stress field, prior to stimulation, was characterized by strong variations resulting from the presence of natural fractures and facies in the study area. The stress field was demonstrated to vary from the regional stress in the vicinity of the fractures and facies and showed regions with high and low stress anisotropy.

Stress anisotropy is a key initial stress condition required by the completion engineers to successfully plan engineered stages targeting geological and geomechanical sweet spots, with low stress anisotropy.

Copyright © American Institute of Chemical Engineers. All rights reserved.

INJECTION-INDUCED RESERVOIR DAMAGE AND ITS EFFECT ON EARLY STAGE PRODUCTION

Andrey Filippov, Xinli Jia, and Kim Thornton, Halliburton

Abstract

This paper presents a method for the fast evaluation of fracture-stimulated reservoir production. For the calculation of injection-induced formation damage during fracturing and its effect on production decline, an efficient numerical model with a three-phase transient analysis of pressure distribution was built and validated using the predictions from reservoir solvers and field data. This model solves for gas-, oil-, and water-flow parameters, and accounts for the gas-oil phase transition. It has been implemented in a numerical code and compared with predictions from commercial software and available field data, such as production decline curves.

The leak-off of fracturing fluids into the formation affects the size and conductivity of fractures and results in formation flooding in a narrow zone near the fractures. This result can significantly affect the transient depletion of the reservoir, particularly during early production. For an adequate description of the fracture productivity, it is important to model these processes in sequence to demonstrate the formation damage envelope and its effect on the production that follows the fracturing process. All gas, oil, and water phases are accounted for in the model, together with their relative permeabilities change and the phase transition effects.

The developed numerical model for hydrocarbon production has been implemented in commercial software and used for the sensitivity analysis of reservoir productivity regarding changes of fracture size and spacing; it can also be used to assess reservoir permeability in the fractured condensate reservoirs and account for multiphase reservoir flows and reservoir properties (Filippov et al. 2016). In this paper, the model is used to simulate the effect of formation damage attributable to leak-off during the fracturing process, resulting in a dramatic change of the production decline curves at early stages of production.

Introduction

Gas production from fractured reservoirs can be very prolific, despite the extremely low matrix permeability that may be encountered. Over the last several decades, hydraulic fracturing has been successfully used for unconventional oil and gas recovery. The fracturing process is implemented by injecting fracturing fluid through a perforation at a sufficiently high rate and pressure to break open the productive formation. The proppant materials, which are used to keep the fractures open after injection, are also injected with the fluid. After completing the injection process, the proppant particles will remain inside the fracture. The fluids are primarily pumped back to the surface, but partially leak into the nearby formation rocks, forming what is known as the formation damage envelope. Consequently, there are two major problems following hydraulic fracturing: the estimated size of the fractures in terms of effective length, and the evaluation of the near-fracture formation damage in terms of reservoir permeability change and phase distribution and the size of the envelope. In this study, the latter problem is considered. The fluids injection is a highly transient process, with the complicated effects of multiphase flows inside the reservoir. A lack of knowledge and means to estimate the effects will result in the over-prediction of the production rate. Rigorous modeling and predictions of the effects on the reservoir during hydraulic fracturing and production are necessary for designing the fracturing process efficiently and optimizing the hydrocarbon recovery rate.

The traditional approach to productivity assessments of fractured reservoirs includes using a detailed 3D reservoir model, with a detailed fracture network geometry obtained from separate mechanical modeling and/or field observations using microseismic devices, and simplified analytical and semi-analytical models. Although the detailed 3D analysis is both time- and resource-consuming and requires highly qualified personnel to run, the simplified analytical equations are not applicable for evaluating the condensate field production.

It is possible to develop a model with simplified geometry, but with detailed accounting for multiphase dynamics that will quickly predict pressure depletion and production decline of fracture-stimulated condensate reservoirs. The effectiveness of such a model can enable it to be used as an application to existing wellbore simulators, reducing the need to combine them with a detailed 3D reservoir simulator when a fast parametric or sensitivity analysis is needed. It is obvious that more than one of these types of models can be developed based on different physical and numerical approaches. This paper shows how using this class of models for the fast and accurate economic analysis of fracture-stimulated production from condensate reservoirs can be efficiently applied to determine optimized fracture parameters, including the number of created fractures, their length and spacing, and production time.

As a proof of concept of this approach, an efficient numerical model was created by Filippov et al. (2016). This model is based on simplified parallel-planes geometry using three-phase flow transient analysis of pressure distribution in an extended stimulated domain, and accounts for gas condensation and water flow. Because the model enables phase transition between oil and gas, the banking effect (production reduction attributable to gas condensation) was identified and analyzed. This numerical model has been implemented in MATLAB® software code and was used in this study to observe the detrimental effect of leaked fluids on fracture productivity at early production stages.

The Near-Reservoir Model

In a simple model optimized for quick analysis, fractures are assumed to be planar, parallel to one another, and identical, with length, *L*, and height, *H*, and spacing, *2d*, between two neighboring fractures, as shown in **Fig. 1**. To simulate production from the condensate reservoirs, the mathematical model includes the mass-balance equations for condensate (oil and gas phases) and water.

$$\frac{\partial \rho_\alpha S_\alpha \varphi}{\partial t} + div\left(\rho_\alpha \boldsymbol{u}_\alpha\right) = q_\alpha \qquad \text{(1)}$$

$$\boldsymbol{u}_\alpha = -\frac{k_\alpha}{\mu_\alpha}\nabla p_\alpha; \qquad \alpha = g, o$$

$$z = 0, \quad x < L_t: \quad p = p_f(t, x)$$

$$z = 0, \quad x > L_t: \quad \frac{\partial p}{\partial z} = 0 \qquad \text{(2)}$$

$$z = d: \quad \frac{\partial p}{\partial x} = 0; \qquad x \to \infty: \quad p \to P_r$$

where α is the index of the component (oil, gas, water), S_a is component saturation, j is the formation porosity, r_α is the component density, k_α and μ_α are the component's permeability and viscosity, respectively, P_r is the reservoir pressure, q_α is the generation rate for the α -th component owing to the phase transition and L_t is the fracture length. The value z = 0 corresponds to the fracture plane. The pressure in fracture p_f is found from the auxiliary problem:

$$
\begin{aligned}
z=0: &\quad \frac{\partial}{\partial x}\sum_{\alpha}\left(\rho_{\alpha}\frac{k_{f\alpha}}{\mu_{\alpha}}\right)\frac{\partial p_f}{\partial x}=\left(\sum_{\alpha}\frac{k_{\alpha}}{\mu_{\alpha}}\right)\frac{\partial p}{\partial x} \\
x=0: &\quad p_f=P_0 \\
x=L_{t:} &\quad \frac{\partial p_f}{\partial x}=0
\end{aligned}
\qquad (3)
$$

where k_{fa} is the permeability of a phase in the fracture and P_0 is the pressure at the beginning of the fracture. The solutions of equations 1 through 3 are matched iteratively at each time step. The condensate-gas equilibrium ratio is to be calculated using the PVT data tables, depending on the composition of the hydrocarbon phase in the formation.

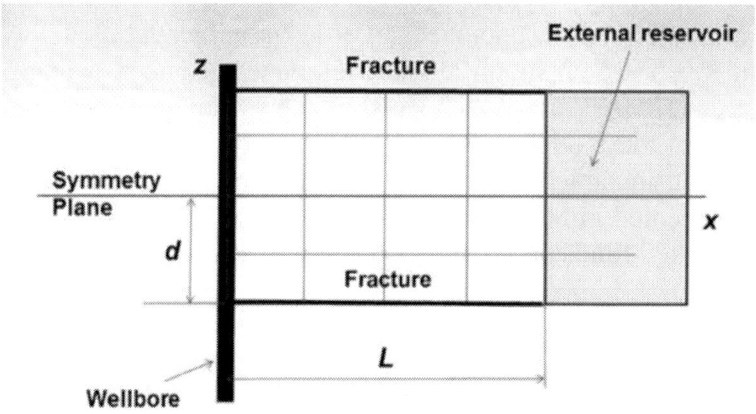

Fig. 1. Geometry of the model.

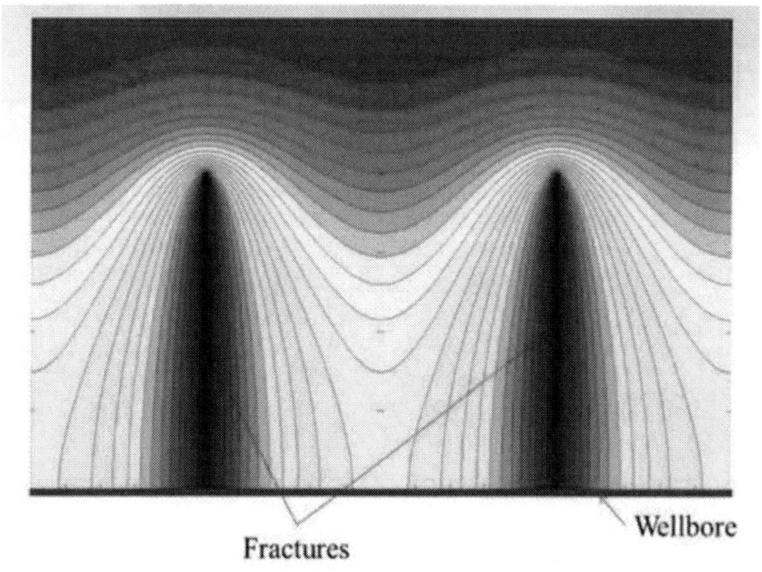

Fig. 2. Isobars near the fractures.

Transient equations 1 through 3 have been discretized using finite differences (Aziz and Settari 1979) and solved using the Alternating Directions Implicit (ADI) method. As an example, **Fig. 2** shows a calculated pressure distribution in the vicinity of two fractures.

The suggested combination of simple, but representative, geometry and detailed accounting for mass transport and physical properties enables fast and rigorous calculations of production depletion in the wellbore and phase transition in the near-well reservoir. This includes the prediction of the banking effect, which occurs when the condensation of liquid from gas results in a significant decrease in the production rate.

For modeling the transient pressure and phase concentration fields during fracturing, the current fracture length L_t is assumed to linearly depend on time t:

$$L_t = L\min\left(1, \frac{t}{t_f}\right) \qquad (4)$$

where t_f is time of fracturing. The boundary condition on the fracture has the form

$$\begin{aligned} &z = 0, \quad x < L_t: \quad p = p_f(t,x) \\ &z = 0, \quad x > L_t: \quad \frac{\partial p}{\partial z} = 0 \\ &z = d: \quad \frac{\partial p}{\partial x} = 0 \\ &x \to \infty: \quad p \to P_r \end{aligned} \qquad (5)$$

The test simulations using the described model have been performed using parameters of water as fracturing liquid.

Modeling the Injection and Production Processes

The injection process begins when the fracture origin pressure P_0 exceeds the reservoir pressure P_r, so that water begins to penetrate the pores of the formation matrix. **Fig. 3** shows the water envelope surrounding the propagating fracture, which grows both along the fracture and in the direction normal to it. The water saturation in the envelope is the maximal saturation that can be reached for the given matrix; consequently, the permeabilities of gas and oil in this zone are very small. Simultaneously growing is the high pressure area around the fracture, which expands much farther into the inter-fracture domain.

If at moment $t = t_f$ the pressure difference $Po - Pr$ changes sign and becomes negative, the fracture begins to produce. The described higher pressure zone between the fractures (**Fig. 4**) remains pronounced for a while, slowly dissipating. **Fig. 5** shows the pressure profile near the fracture at a late stage of production.

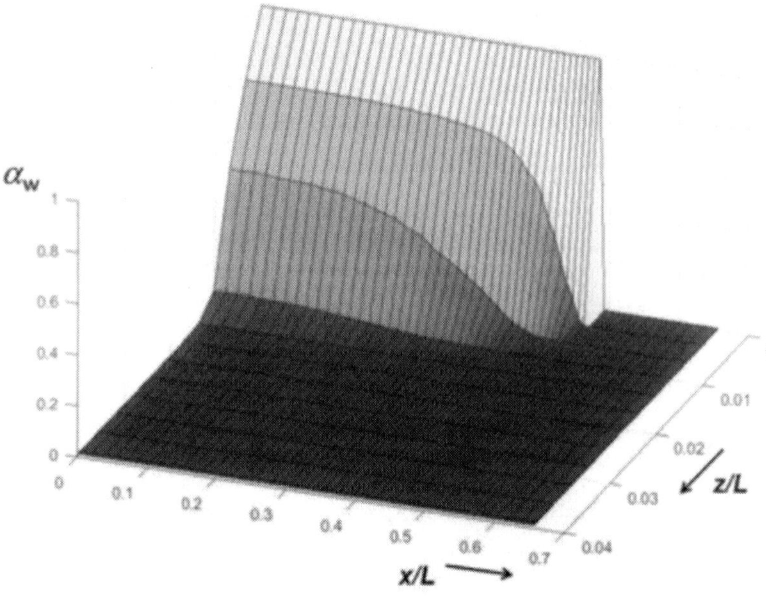

Fig. 3. Water volume fraction distribution around growing fracture.

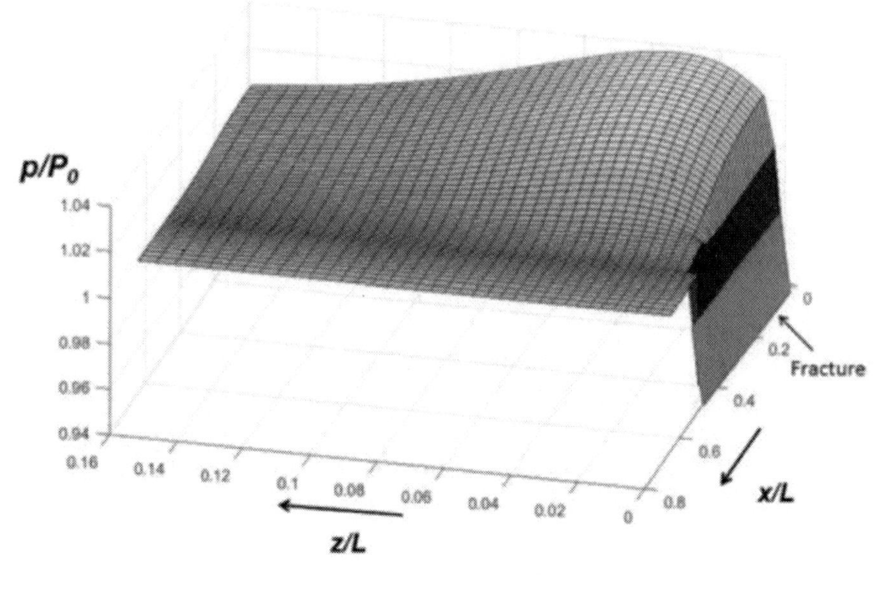

Fig. 4. Pressure distribution near the fracture soon after the beginning of production.

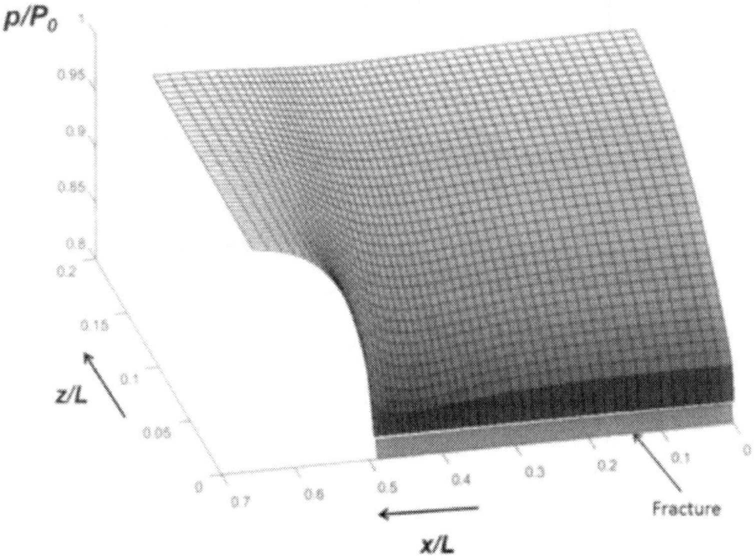

Fig. 5. Pressure distribution near the fracture at a later stage of production.

Changes of oil and gas permeability and pressure distribution near the fracture results in significant changes of predicted production rates, particularly at the early stages of production. A typical production decline curve, calculated using the described model, but without the injection stage, is shown by red line at **Fig. 6**. The predicted production rates at the first days of production are so high that they are typically neglected when comparing the modeled and the field data, or when performing the history matching analysis. Accounting for the water injection yields substantially lower and much more realistic production rates, shown by blue line in Fig. 6, at the beginning of production. Using the described model can be helpful in practical applications, enabling a comparison with measured production rates beginning with the first days of operation.

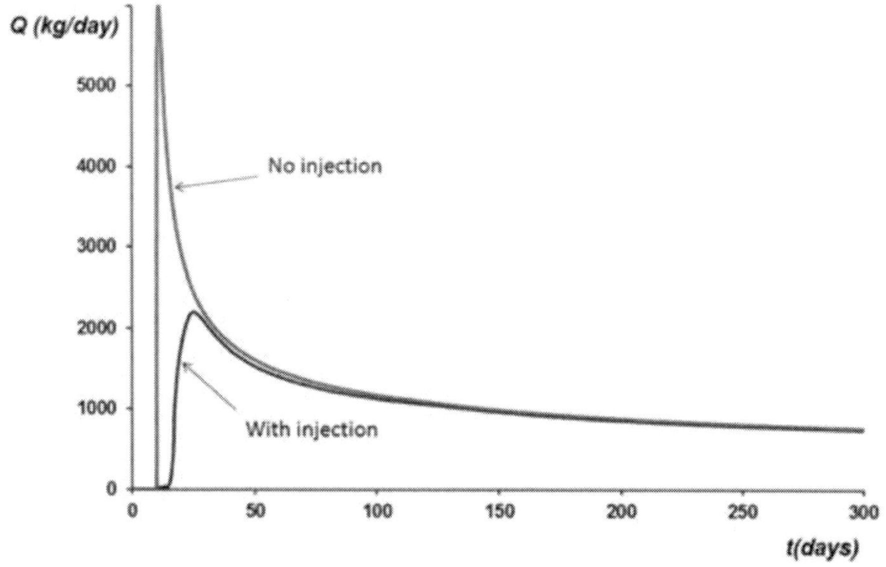

Fig. 6. Production decline curves, calculated with account for water injection (blue line) and without injection stage (red line).

Conclusions

For efficient and rigorous numerical analysis of production in fracture-stimulated reservoirs, an approach has been developed that combines a simplified fracture-wellbore geometry and detailed three-phase transient simulation of pressure distribution in an extended stimulated domain, while accounting for gas condensation and water flow. The resulting model enables the phase transition between oil and gas and accounts for the banking effect (production reduction is attributable to gas condensation). The model has been shown to enable the simulation of the formation of a water envelope around the fracture as a result of leak-off and water injection during fracturing, leading to more realistic shapes of the production decline curves at the early stages of production.

The model has been implemented in MATLAB software code and can be easily used for the sensitivity analysis of reservoir productivity regarding the water damage, changes in fracture size and spacing, as well as reservoir permeability.

References

1. Aziz, K. and A. Settari (1979). *Petroleum Reservoir Simulation*. London, United Kingdom: Applied Science Publishers, Ltd.
2. Filippov, A., X. Jia, X., T. McNealy, and V. Khoriakov, (2016). Fast Economic Analysis and Optimization of Fracture-Stimulated Wells in Condensate Reservoirs. Presented at SPE/IAEE Hydrocarbon Economics and Evaluation Symposium, Houston, Texas, 17–18 May. Paper SPE-179957-MS.

445345 On the Mechanisms of Sequenced Fracturing

Abstract:

Fracturing over long horizontal intervals while using only limited-entry techniques for diversion leaves areas along the wellbore understimulated. Improving wellbore coverage is addressed by using diverters, often in the form of particulates designed to temporarily plug the formation and direct fluids away from the stimulated areas. The success of the wellbore coverage and the overall stimulation are dependent on the ability to formulate a diverter that is effective at plugging, and that can be transported from surface to downhole with little dispersion. The considerations for successful diversion and overall fracturing treatment are integrated in a technique called sequenced fracturing.

Sequenced fracturing was studied with the objective of understanding how diverters are transported and how they plug fractures and generate effective diversion. Based on a review and experiments, methods to prevent dispersion of the particles could be compared and mechanisms of diverter failures investigated. Simulations were used to investigate the effects of diversion on the pressure response and overall quality of the stimulation.

Results from laboratory experiments, computer simulations, and well performance evaluations were analyzed collectively to reach three main conclusions. First, particulate diverters are effective at plugging fractures provided they can be transported from surface to downhole with minimal dispersion. The use of fiber is instrumental in minimizing the adverse effects of wellbore dispersion on diversion effectiveness. Second, effective diversion should rely on plugging the near-wellbore area. Computer simulations show that diverter misplaced in the fracture, at an excessive distance from the wellbore, is more likely to fail under differential pressure. Third, computer simulations showed that pressure response and impact on stimulation are functions of the rock and the well completion. Examples are provided to illustrate how in-situ stress configuration affects pressure response and stimulation effectiveness of a sequenced fracturing treatment.

Copyright © American Institute of Chemical Engineers. All rights reserved.

445346 Generating Heterogeneous Proppant Placement in Hydraulic Fracturing by In-Situ Channelization

Abstract:

The primary goal of a hydraulic fracturing treatment is to create a highly conductive flow path from the formation to the wellbore to increase well production. Engineering a fracture with high conductivity and half-length has been a source of continuous effort and challenge over the years. Because proppant-pack retained permeability is often a small fraction of the maximum expected value, techniques have been developed to create networks of open channels inside the fracture. An in-situ channelization (ISC) process generates heterogeneous proppant placement that leads to highly conductive channels.

ISC is unique to fiber-based composite fluids, by which channels spontaneously form in a fracture before fracture closure. The ISC process involves pumping a composite fluid that segregates into proppant pillars. This type of heterogeneous proppant placement is facilitated by the presence of fibers that hinder proppant settling and anchor the proppant pillars in place. One benefit of such process is the resulting high conductivity due to the channels, which are maintained open by pillars made of fibers and proppants. ISC has been observed and studied in laboratory conditions. It occurs with linear gels and crosslinked fluids within a well-defined range of parameters (fiber loading, proppant loading, fracture width, and rheology). With crosslinked fluids, ISC can be triggered by breaking the carrier fluid shortly before the fracture closure. With linear gels, ISC forms spontaneously after the placement of the composite fluid in the fracture. Experimental evidence also supports that pillars formed by ISC do not spread under closure pressure, but remain contained and competent.

The study of ISC required unique experimental setups of sufficiently large scale to capture the dimensions of a real fracture. The experimental evidence enabled identification of the mechanisms of ISC formation, the factors that favor the formation of channels, and the impact of ISC on fracture properties.

Copyright © American Institute of Chemical Engineers. All rights reserved.

16AIChE
Spring Meeting, Houston, TX

Development of an Environmental Friendly Pseudo-Oil Based Fluid for Shale Stability

Yidi Wang
The University of Tulsa
Yidi-wang@utulsa.edu

Feifei Zhang
The University of Tulsa
Feifei-zhang@utulsa.edu

Prepared for Presentation at
American Institute of Chemical Engineers
2016 Spring Meeting
11th Global Congress on Process Safety
Houston, Texas
April 10-14, 2016

UNPUBLISHED

AIChE shall not be responsible for statements or opinions contained
in papers or printed in its publications

Development of an Environmental Friendly Pseudo-Oil Based Fluid for Shale Stability

Yidi Wang
The University of Tulsa
Yidi-wang@utulsa.edu

Feifei Zhang
The University of Tulsa
Feifei-zhang@utulsa.edu

Keywords: Cationic alkyl ploygIycoside, synthesis, technique, drilling fluid, evaluation, Pseudo-oil based drilling fluid, development

Abstract

Shale instability is one of the major concerns for unconventional exploration and production. Conventional water-based fluid (WBM) system performs poorly in shale formations and causes serious shale stability problems. Oil-based fluid are commonly used in shale drilling, which have serious environmental issues and huge cost compared to WBM. Development of a Pseudo-Oil Based Fluid, which is water based but has the properties to stabilize shale formation as OBM, is attractive to shale drilling from both environmental and economic perspectives.

The main source of shale instability is the change of stress state and shale strength through shale-fluid interaction. The components of shale include water sensitive materials, which increase volumes when exposed to aqueous environments. With conventional water-base drilling fluids, shale formations may swell and slough as a result of interaction with the drilling fluid. Furthermore, the shale disperses into colloidal sized particles of clay has an adverse effect on drilling fluids. Inhibiting shale-water interaction during drilling can increase the wellbore stability.

In this paper, a pseudo-oil based fluid is developed by using a cationic alkyl ploygIycoside (CAPG) surfactant that is from natural agricultural products, aliphatic alcohols, and amine reagents. The CAPG is particularly effective as a shale inhibitor; it creates a protection coating at the surface of the shale and isolates the shale. Thus, the water-shale interaction is significantly inhabited. Since CAPG is synthesized by using natural products, and it is completely environmental harmless and can decompose quickly after the drilling completed.

The pseudo-oil based drilling fluid is tested under field conditions and compared to the original oil-based fluid. A system including K-PAM as viscosifier, JA-B as Polyamines inhibitor and CN as temperature resistance agent is developed to be used with CAPG to get the desired fluid properties. Results show that the two fluids have almost the same rheology and lubricity, and slightly lower inhibitive ability. This work provides an environmental friendly and economic approach to stabilize shale formations.

Key words: Cationic alkyl ploygIycoside, synthesis, technique, drilling fluid, evaluation, Pseudo-oil based drilling fluid, development

1. Introduction

In the past two decades, shale oil and gas production has become a significant part of the hydrocarbon production. One of the most challenging part of the shale oil/gas exploration and production is to drill the well successfully in the unstable shale formation. The traditional approach to drill through the shale formation is to use the oil based mud (OBM). The environmental issues and high cost makes it more and more urgent to develop a water-based mud (WBM), which also has similar properties to stabilize the shale formation. Cationic alkyl glucoside (CAPG) is a type of green non-ionic surfactants, it is a mild non-ionic and cationic surfactants which contains glucose, fatty alcohols, and amino. The CAPG can create a protection coating at the surface of the shale and isolates the shale, which makes it a good candidate for shale inhibitor. In recent years, a number of CAPG products have been developed in the field of drilling fluid [1, 2]. In this paper, we use a CAPG surfactant, which has similar properties with the oil-based fluids, develops a pseudo-oil based fluid.

2. Fluid Additive Selection

The requirements for a pseudo-oil based drilling fluid system mainly include three aspects: excellent lubrication, strong inhibitory, and resistance to high temperature. To satisfy these conditions, glycosides, polyamines and temperature stabilizer is selected as additional additives.

Based on large number of experimental tests, K-PAM is chosen as anti-collapsed agent, stick agent, and effective agent to inhibit shale swelling. Due to the long carbon chain of alkyl in the cation alkyl glycosides, the selected additive has excellent of lubrication. The hydroxyl in the additive can adsorb on the clay and form an isolation layer on the shale, which gives K-PAM strong inhibitive capacity and good lubricative property. And by adding temperature stability SN, system stability can improve under high temperature. The main additives selected for the pseudo-oil drilling fluids are:

Cationic alkyl glycoside: CAPG-1

Polyamine inhibitor: JA-B

Polymer viscosifier: K-PAM

Temperature stability: SN

Through a large number of test and analysis, a preliminary recipe for the pseudo-oil based drilling fluids is obtained, which can be shown as:

Table.1 Preliminary Recipe for Pseudo-oil Based Drilling Fluids

Bentonite	NaOH	Na2CO3	K-PAM	LV-CMC	SMP-1	SPNH	CAPG-1	JA-B	SN	KCl
3-4%	0.15-0.2%	0.15%	0.08-0.5%	1.2-1.5%	3%	2%	2-3%	1-2%	0.2%	1%

3. Comparison with Oil Base Fluid

Experiments were conducted to compare the basic performances and inhibition ability of pseudo-oil based drilling fluids and oil-based drilling fluids. A commonly used oil-based fluid for shale drilling is selected for comparison purpose. The recipe of the oil-based fluid is:

Table.2 Preliminary Recipe for Oil-based Drilling Fluid

Diesel	Viscosifier	Major emulsifying agent	Complementary emulsifying agent	Wetting agent	CaO
320ml	0.5%	3%	2%	1%	2%

The basic parameters of the two fluids are shown in Table.3. Inhibition performance and lubricity performance comparisons results can be seen in Table.4 and Table.5.

In Table.3, PV is plastic viscosity, YP is yield point, and ES is electrical stability. And API FL is fluid filtration loss volume at ambient temperature and 100 psi differential pressure for 30 min, while HTHP is fluid filter loss volume at 300°F and 500 psi differential pressure for 30 min.

Table.3 Basic Parameters of Drilling Fluids

Sample	PV mpa.s	YP Pa	YP/PV $10^{-3}s^{-1}$	API FL ml	HTHP ml	ES V
Oil-based fluid	42.0	23.0	0.55	1.7	5.2	897
Pseudo-oil based fluid	54.0	16.4	0.43	3.4	6.0	

From above table, oil-based fluid has lower plastic viscosity than pseudo-oil based fluid, indicating that oil-based fluid is more capable of drilling rapidly. And oil-based fluid has higher yield point than pseudo-oil based fluid, which implies that oil-based fluid has better ability to suspend cuttings. While YP/PV ratios are near, we can see both fluids have similar good ability to provide cutting transport. In addition, despite the fact that API FL of oil-based fluid is half of the API of pseudo-oil based fluid, they perform nearly well under high temperature and high pressure.

Table.4 Inhibition Performance Comparisons of Drilling Fluids

Drilling Fluid	Water	Pseudo-oil based fluid	Oil-based fluid
Hot-rolling recovery rate (%)	1.23	85.47	93.76

From above table, oil-based fluid has higher recovery rate than pseudo-oil fluid. Shale hot-rolling recovery rate can be used to represent inhibition dispersion capacity of drilling fluid system to shale. Hot-rolling recovery experiment results indicate that pseudo-oil based fluid can inhibit clay dispersion and perform good stability in shale as oil- based fluid.

In Table.5, Kf is lubrication coefficient. All data is obtained from Lubricity Tester.

Table.5 Lubricity Performance Comparisons of Drilling Fluids

Lubricity	10'	Water/5'	Kf
Oil-based fluid	10.4	30.4	0.116
Pseudo-oil based fluid	12.9	40.4	0.109

From above table, lubricity performance of pseudo-oil based oil is better than oil-based fluid.

In conclusion, the basic properties and inhibition property of pseudo-oil based fluids are a litter worse than oil-based drilling fluids. But lubricity of pseudo-oil based fluids is better. With the temperature resistance, pseudo-oil based fluids can obtain high-temperature resistance ability until 160 °C from experimental results.

4. Discussion on inhibitive Mechanism of Pseudo-oil based Fluid System

In pseudo-oil based fluid system, the CAPG-1-octyl hydroxypropyl glycosidase trimethyl ammonium chloride and a strong polyamine inhibitor with silicon functional group and temperature stability SN are introduced to inhibit clay expansion in the shale formation.

From the lab infrared spectroscopic analysis, we roughly obtain the molecular structure of CAPG-1. It can be seen in Figure.1.

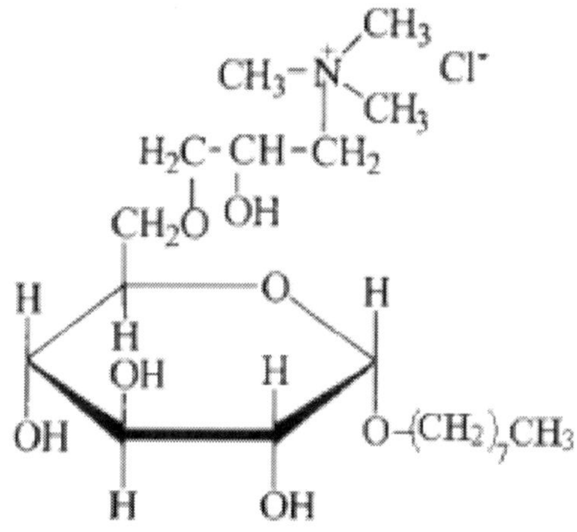

Figure.1 CAPG-1 molecular structure

CAPG is a cationic surface-active agent; it can form a semipermeable membrane that is similar to oil based fluid so that it has good inhibition, lubricity, and reservoir protecting abilities. There are 4 hydrophilic hydroxyl groups (-OH), which can adsorb on well wall and drilled cuttings surface. Unlike the traditional water based mud, the CAPG fluid can dehydrate the formation rock once it entering the formation. The oil wet oleophilic alkoxy group (-O $(CH_2)_nCH_3$) in the CAPG grow towards the outside of the molecular (as shown in Figure 1), which forms a water-phobic membrane like the oil based drilling fluids. This membrane can separate the water in formation rock from the water in the drilling fluid. The change of permeation affect the interaction between water and formation rock from two aspects: one is to increase viscosity of mud filtrate, decreases permeability, decrease filtration loss; the other aspect is to decrease water activity in drilling fluid, decrease pore pressure, decrease the amount of water going to shale. The cationic amino in CAPG and the cationic amino in polyamine can attach to the surface of clay particle by electrostatic interaction, meanwhile, the nitrogen atoms in ammonium ion forms hydrogen-bond interaction with clay crystal layer surface. The combination of these two factors is stronger than the hydrogen bond between water and crystal layer surface. Thus, the some water molecular between the clay layers can be squeezed out, which decreases the interplanar space. Furthermore, polyamine can bound the clay layer together, which decreases the hydration expansion of clay.

The polyamine used in the pseudo-oil based fluid is a modified polyamine agent by adding silicious functional group, which can further inhibit clay hydration. There are a large number of hydrogen bonds in this polyamine, which can react with water and turn free water in the system into absorbed water or irreducible water. In addition, the membrane also reduces filtration loss.

In the end, by adding temperature stability SN into pseudo-oil based fluid system, free radical generation is inhabited, so the system stability under high temperature is improved.

5. Conclusion

In this paper, pseudo-oil based drilling fluid with strong inhibitive ability and lubricity is developed. The mechanism, which causes the strong inhibitive ability for this fluid, is investigated. The conclusions and contributions of this thesis include:

1. The effect of CAPG-1 on inhibitive ability, lubricity and other basic properties of drilling fluids are evaluated for water, a bentonite fluid and a simulated field fluid. The test results show that CAPG-1 has strong inhibitive ability, good lubricity, and filtration properties. In the water, the cuttings collection ratio is more than 99%. In the bentonite fluid, the inhibitive ratio for apparent viscosity is more than 95% in ambient condition.
2. In several evaluation systems, the CAPG-1 has good lubricity and strong inhibitive ability. It has little effect on the basic properties of the drilling fluid. As the CAPG-1 concentration increases, the performance of the fluids improves.
3. By using CAPG-1 as lubricity inhibitor, pseudo-oil based drilling fluid with strong inhibitive ability is developed. A system includes K-PAM as viscosifier, JA-B as inhibitor and CN as temperature resistance agent, can be used with CAPG-1 to get the pseudo-oil based fluid properties.
4. Compared to the oil based drilling fluid, the pseudo-oil based drilling fluid has almost the same rheology, slightly higher water loss, and slightly lower inhibitive ability. The rotation recollection rate of the pseudo-oil based drilling fluid is about 85%, which is slightly lower than the oil based drilling fluid's 90%.

Reference

[1] Clapper D K, Watson S K. Shale Stabilizing Drilling Fluid Employing Saccharide Derivatives, EP[J]. Pat, 1996, 702073.

[2] ConocoPhillips Company.Surfactant system to increase hydrocarbon[P] ,EP 2734705 A1 2014-05-28 .

[3] Chan A F, Bohon W M, Blumer D J, et al. An alkyl polyglycoside, an ethoxylated alcohol, a caustic and an alkyl alcohol: U.S. Patent 6,130,199[P]. 2000-10-10.

[4] Chan A F. Well cleanout using caustic alkyl polyglycoside compositions: U.S. Patent 5,374,361[P]. 1994-12-20.

[5] Chan A F. Acidic surfactant composition and method for cleaning wellbore and flowline surfaces using the surfactant composition: U.S. Patent 5,977,032[P]. 1999-11-2.

[6] Chan A F, Bohon W M, Blumer D J, et al. Surfactant composition and method for cleaning

wellbore and oil field surfaces using the surfactant composition: U.S. Patent 5,996,692[P]. 1999-12-7.

[7] Chan A F, Ly K T. Surfactant blends for well operations: U.S. Patent 5,830,831[P]. 1998-11-3.

[8] Claus Peter Herold. Verwendung oberflächenaktiver alkylglycosidverbindungen in wasser-und öl-basierten bohrspülungen und anderen bohrlochbehandlungsmitteln [P] ,WO1992002594 A1， 1992-02-20.

[9] Michele Scott Albrecht. Silica-based drilling mud comprising glycoside lubricants with amino-linked alkyl chains [P], US7320951 B2, 2008-1-20.

[10] Si X, Wang Z, Wei J, et al. Study on synthesis of cationic methyl glucoside which was used in drilling fluid[C]//ABSTRACTS OF PAPERS OF THE AMERICAN CHEMICAL SOCIETY. 1155 16TH ST, NW, WASHINGTON, DC 20036 USA: AMER CHEMICAL SOC, 2012, 243.

[11] Si X, Wang Z, Wei J, et al. Performance test for cationic methyl polyglucoside which was used in drilling fluid[C]//ABSTRACTS OF PAPERS OF THE AMERICAN CHEMICAL SOCIETY. 1155 16TH ST, NW, WASHINGTON, DC 20036 USA: AMER CHEMICAL SOC, 2012, 243.

[12] Si X, Wang Z, Wei J, et al. Influence of catalyst on the synthesis of cationic alkyl glucoside[C]//ABSTRACTS OF PAPERS OF THE AMERICAN CHEMICAL SOCIETY. 1155 16TH ST, NW, WASHINGTON, DC 20036 USA: AMER CHEMICAL SOC, 2013, 246.

[13] Gao P, Towler B F, Jiang H. Feasibility Investigation of CO2 Miscible Flooding in South Slattery Minnelusa Reservoir, Wyoming[C]//SPE Western Regional Meeting. Society of Petroleum Engineers, 2010.

[14] Wang B, Wang Z, Sun Y, et al. The study of formulation and performance of formate drilling and completion fluid system[J]. Natural Science, 2013, 5(09): 997.

Using a Large Data Bank to Develop a Robust and Simple Model to Estimate the Saturation Pressure of Oil Reservoirs

By

Hassan Alkandari[1], Adel Elsharkawy[1], Adel Malallah, Osamah Alomair

Petroleum Engineering Department, College of Engineering & Petroleum
Kuwait University
Po Box 5969 Safat 13060-Kuwait

Abstract

Measurements of saturation pressure are crucial for all hydrocarbon reservoir fluids. Below the crude oil saturation pressure gas reaches a critical saturation; then, a two phase flow occurs and results in decreasing oil production and recovery. To maximize oil production and recovery, the reservoir pressure has to be maintained closer to the original saturation pressure. This pressure is normally measured using bottom-hole samples or surface recombined samples of oil and gas. Occasionally, the samples become unavailable and the pressure needs to be estimated using computational methods.

In this study, a large data bank of 231 crude oil compositions and saturation pressure measurements including literature data, unpublished data, and newly measured data were used to develop two empirical models to predict the saturation pressure of a variety of crude oils if the oil sample is unavailable or the experimental measurements are unreliable. The first proposed model utilizes the extended compositions of hydrocarbons up to the heptane plus fraction in addition to non-hydrocarbons. The second model uses the lumped compositions of light, intermediate, and heavy components in addition to non-hydrocarbon components as the input variables. The lumped model has the advantage of using fewer input parameters while maintaining the thermodynamics. The accuracy and validity of both models to calculate the saturation pressure for volatile oils, black oils, and heavy oils are presented using several compositional data. The models performance is also compared to the Peng-Robinson equation-of-state (PR-EOS), and the Soave-Redlich-Kwong equation-of-state (SRK-EOS) as well as all published methods that use compositions as input variables. The comparison indicates that the proposed models are much simpler and more accurate than the other computational methods. The proposed models treat the heptane-plus fraction as a single component, thus eliminating the splitting and characterizing the plus fractions and the binary interaction parameters needed for the EOS's calculations.

Key words: Saturation pressure (Ps), Bubble-point pressure (BPP), Equation of state (EOS), Correlations, and PVT Data

[1] Corresponding Authors: Hassan Alkandari , Hassan@hkandari.com, Adel Elsharkawy, adel.elsharkawy@ku.edu.kw

Introduction:

Reservoir fluid properties are an essential source of information for all calculations on hydrocarbon reserves. Accurate estimation of the pressure-volume-temperature (PVT) measurement is crucial. Of all the PVT properties, saturation pressure (Ps) is considered the most important parameter. It is well known that gas starts to flow once the reservoir pressure drops below the bubble point pressure and the gas saturation reaches a critical value. To minimize the gas flow and maximize the oil recovery, it is very important to maintain the reservoir pressure close to the original bubble point pressure.

Although the PVT properties are experimentally measured in laboratories and provide reliable results, this process is expensive, time consuming, and sometimes unavailable; therefore, measured field data (separator gas oil ratio, separator pressure, stock-tank oil specific gravity, and reservoir temperature) or reservoir fluid compositions are used to estimate the bubble point pressure.

There are two methods in the petroleum industry to calculate the saturation pressure of crude oil when experimental is unavailable; either using production data or compositional data.

Methods Based on Production Data (Composition is unavailable)

Numerous attempts have been made since the 1940s to develop correlations to estimate the saturation pressure from production data, such as producing gas oil ratio, oil gravity, gas gravity, reservoir pressure and temperature. Table 1 presents summary of all the methods that use production data to estimate the saturation pressure.

Methods based on Compositional data

Many authors have reported that the application of any correlation in other regions results in serious errors. These conclusions arise from the fact that regional crude oils represent a physical-chemical trend; therefore, all attempts to develop a general correlation using field measured parameters to estimate the saturation pressure from a large data bank have failed. For this reason, Elsharkawy (2003) made the first attempt to correlate the saturation pressure to the reservoir fluid composition instead of using the production data. In recent years, there have been several attempts to estimate the saturation pressure of oil reservoirs. These methods are reported in Table 1.

Elsharkawy (2003) presented a correlation of the bubble point pressure for the oil reservoir using the same input data needed for the equation-of-state (EOS) calculations, i.e., the compositional data and the reservoir temperature. Elsharkawy's model is based on the saturation pressure measurements of 60 crude oil samples from the Middle East and 75 collected from the literature. He demonstrates that the calculation of the saturation pressure depends on the type of EOS, the number of pseudo components, the method used to characterize the pseudo-components and the binary interaction number. Elsharkawy compared the performance of his model to the SRK-EOS and PR-EOS for black oils, volatile oils, heavy oils, and synthetic oils. He indicated that his model eliminates splitting the heavy plus fraction, characterizing the pseudo-components and inclusion of the binary interaction parameters needed for the EOS calculations.

Later, Tale et al. (2009) made the second attempt to model the saturation pressure of crude oil using 132 data samples published by Elsharkawy (2003). Tale's model, which is based on the alternating conditional expectation (ACE), has a set of 14 equations with 42 constants. The average absolute relative error (AARE) reported by Tale et al. indicates that the model did not perform well with the selected data. The ACE model has an AARE of 12.3% compared with AARE of 6.7% by Elsharkawy (2003) and 8.1% AARE for the SRK-EOS.

AlQuraishi (2009) attempted to model the saturation pressure using Elsharkawy's (2003) data. The model was developed using linear genetic programming. He correlated the saturation pressure to the formation temperature, methane content, and heptanes plus molecular weight. Thus, AlQuraishi's model neglected the effect of the intermediate and heptane plus content on the saturation pressure. The error analysis reported by AlQuraishi indicates that his model has a 5.8% error compared to 7.7% by Elsharkawy, 9.9% using the SRK-EOS, and 10.1% using the PR-EOS.

Bandyopadhyay and Sharma (2011) presented a model to predict the bubble point pressure, using temperature and fluid composition. They used 129 experimental and literature data originally published by Elsharkawy (2003) to develop their model. The model introduced a temperature interaction parameter that depends on compositions. The proposed model has a set of 24 coefficients and the estimated bubble point pressure is calculated as the sum of the series calculations. They reported an average absolute error (AAR) of 7.83% using their model compared to 8.3% using Elsharkawy's simple model (2003), 10.15% using SRK-EOS, and 10.58% using the PR-EOS.

Khamehchi et al. (2011) used an intelligent proxy simulator (IPS), which is based on an adaptive neural fuzzy inference system (ANFIS) and a structured artificial neural network (ANN) to predict the bubble point pressure and asphaltene onset pressure. The model uses a reservoir fluid composition and temperature to predict the Ps. It is not clear how the Ps is calculated from the proposed method, and they did not present the accuracy of the proposed method compared to the other published methods.

Farasat et al. (2013) presented a new mathematical approach to calculate the crude oil saturation pressure as a function of the temperature and the reservoir fluid composition. The model developed by Farasat et al used 130 experimental data, which were originally published by Elsharkawy (2003). Farasat used a support vector machine (SVM) to model the saturation pressure. They did not clearly demonstrate how to calculate the saturation pressure from the SVM model. Moreover, they did not indicate the overall average absolute deviation (AAD) for their model compared to other published models and both the PR-EOS and SRK-EOS for the 130 data.

Ahmadi et al. (2014) also presented an ANN technique called gene expression programming (GEP) to determine the bubble point pressure of the oil samples. Similar to Farasat et al. (2013), Ahmadi's model used most of the data previously published by Elsharkawy (2003) to construct a non-linear equation using reservoir fluid compositions and temperatures as input parameters. The proposed model is highly complicated and consists of 8 equations that have several constants compared to the one simple equation proposed by Elsharkawy (2003). They compared the performance of their proposed model to the SRK-EOS and PR-EOS; however, their paper does not mention how the heptanes plus fraction were treated and characterized for the EOS calculations. They reported an AAD of 4% using the proposed model compared to 8% using Elsharkawy's model (2003), 10.9% by PR-EOS, 10.5% by SRK-EOS, and 15% by Bandyopadhyay and Sharma (2011).

Gholami et al. (2014) estimated the bubble point pressure using a support vector regression SVR and a supervised learning algorithm SLT based on the statistical learning theory, and then the ACE was used. The model used reservoir fluid compositions and temperature to estimate the bubble point pressure. Gholami et al. used the same data originally published by Elsharkawy (2003). They presented three attempts based on genetic programming to correlate the bubble point pressure: SVR, ACE, and power-law committee machine (PLCM). The ACE-based method resulted in a set of three equations with a matrix of [8 by14] constants. Gholami et al. compared the accuracy of their model to the SRK-EOS and PR-EOS predictions reported by Elsharkawy (2003). Based on their AARD analysis, the PLCM model has the lowest errors followed by the SVR model, Elsharkawy's model, ACE model, PR-EOS, and SRK-EOS.

Lately, Jarrahian et al (2015) presented two empirical models to estimate the saturation pressure of black oils. The first model is based on production information. The second model is based on compositional information mostly from Elsharkawy (2003) and Heidaryan and Moghadasi (2011). They have not presented comparison of the accuracy of their model to Farasat (2013) model, Ahmadi (2014) model, and Gholami (2014) model.

This paper is concerned with the estimation of saturation pressure using compositional data. Therefore, the objective is to develop a robust yet simple model to estimate the saturation pressure for various types of reservoir oils using a larger data bank than that used by other published studies. The accuracy of the proposed models is compared to previously published models as well as SRK-EOS and PR-EOS.

Data bank

This study uses a large data set of 231 crude oil compositions and saturation pressure measurements. The data set consists of 130 data sets originally reported by Elsharkawy (2003); in addition to others collected from literature (Jacopy, 1958; Vogel, 1980; Coats, 1986; Droham, 1980; Dansh, 1992; Moharam, 1996; Rosenegger, 1999); 94 new data sets that have not been utilized before; as well as 7 newly measured samples at the Petroleum Fluid Research Center (PFRC) in Kuwait University. These data represents a wide range of compositions, such as volatile oils, black oils, and heavy oils. The data cover a pressure range of 537 to 5065 psia and a temperature range of 58 to 319 °F, crude oil compositions from C_1 to C_{7+}, non-hydrocarbons, such as N_2, H_2S, and CO_2, and the molecular weight and the specific gravity of the heptane plus fraction.

Data screening and classification

The compositions and the saturation pressure measurements in the data bank are subjected to a meticulous screening process to check the validity of the data. Our screening criteria uses the following steps: the sum of the molar compositions of all the components must add up to one, otherwise, the sample is eliminated. Samples with a missing molecular weight or specific gravity of the plus fraction are also eliminated.

After filtering, the data were reduced to 214 and divided into 179 measurements used to develop the new models, and 35 data sets are used for the blind test of the robustness of the proposed models. Statistical analysis of the input variables of the data set used in training and testing the extended and the lumped models are illustrated in Tables 2.

Several of the reported compositional data have extended analysis up to (C_{20+}). The following techniques are used for lumping or regrouping the extended analysis into single pseudo-fractions (C_{7+}). The mole fraction (X_i) of the individual components of the extended analysis is equal to the plus fraction (Ahmad, 2011).

$$X_{C_{7+}} = \sum_{i=7}^{n} X_i \qquad i=7, 8 \ldots n \qquad (1)$$

The average molecular weight (Mw) of the individual component of the extended analysis is equal to the molecular weight of the plus fraction.

$$Mw_{C_{7+}} = \sum_{i=7}^{n} \frac{X_i Mw_i}{X_{C_{7+}}} \qquad i=7, 8 \ldots n \qquad (2)$$

The density of the total plus-fraction can be calculated as follows:

$$\gamma_{C_{7+}} = \frac{X_{C_{7+}} Mw_{C_{7+}}}{\sum_{i=7}^{n} \frac{X_i Mw_i}{\gamma_i}} \qquad i=7, 8 \ldots n \qquad (3)$$

It is important to note the significance of the density of plus fraction because its density determines its paraffinic naphthenic aromatic (PNA) content.

Equation-of-state calculations

The EOS calculation of the saturation pressure is compared with the newly developed models as well as the other methods. The two EOS calculations considered in this paper are the ones most considered and widely used in the petroleum industry to evaluate the volumetric and phase behavior of reservoir fluid properties: the Soave-Redlich-Kwong EOS (SRK-EOS) and the Peng-Robinson EOS (PR-EOS). It should be noted that the accuracy of any EOS calculation of the saturation pressure largely depends on the number of pseudo-components and the splitting schemes used to divide the heptane plus fraction; as well as the characterization and calculation of the critical properties of the pseudo-components and the binary interaction parameters. Therefore, the two EOS calculations of the saturation pressure are not the same, even though the number of pseudo-components, characterization of the pseudo-components, and binary interaction parameters were the same.

Model development

As previously stated, the objective of this work is to provide the petroleum reservoir engineer with a simple yet accurate tool to estimate the saturation pressure for a variety of crude oils using compositional data when experimental is unavailable. Two models are presented in this study to calculate the saturation pressure of the crude oil samples. The first model uses all input parameters, such as the molar compositions of the hydrocarbon and non-hydrocarbon components similar to the EOS models. This model is referred to as the extended model. The second model uses fewer input parameters in which some of the hydrocarbon components have been lumped into a single group; this is designated as the lumped model. Both models are simpler than all published correlations, physically correct, and have an acceptable accuracy compared with the well-known EOS.

Figure 1 shows a correlation matrix of all the data available under consideration. This figure also indicates the relative importance of each of the input variables into the first proposed model. It agrees with most published correlations and the EOS that the methane content, heptane plus content and temperatures have the greatest impact on the saturation pressure followed by the intermediate content. The data understudy indicates that the non-hydrocarbon has the smallest impact. This result is simply a reflection of the molar contribution of each component comprising the crude oil composition.

Model I - Extended Model

The extended model uses 13 variables comprising the molar compositions of the pure components, heptane plus properties, and reservoir temperatures, which are similar to the input parameters for the EOS. Figure 2 shows the correlation coefficient graph of all the input parameters in the extended model. Thus, the elimination of the intermediate effect has a negative impact on the accuracy of the predictive model. Our proposed extended model has the following form:

$$Ps = 124.72N_2 + 17.57CO_2 + 22.55H_2S + 64.22C_1 - 9.80C_2 - 52.49C_3 + 6.16C_4 - 19.22C_5 - 23.63C_6$$

$$- 21.43C_{7+} + 435.31\gamma_{C7+} + 1.14Mw_{C7+} + 4.29T \quad (4)$$

Where Ps is the saturation pressure in psia, T is the temperature in °F, and the hydrocarbon and non-hydrocarbon compositions are expressed in mole.

Model II - Lumped Model

The second model is much simpler and uses only 7 -variables (the lumped model). Several lumping schemes are adapted in the petroleum industry for phase behavior modeling and reservoir simulation of the petroleum reservoir fluids (Ahmed, 2011). Hydrocarbon components are grouped into a pseudo fraction based on their molecular weight (Whitson, 1980), the similarity of their physical-chemical properties (Lee, 1979), their carbon number, their true boiling point or their number of families based on the chemical structure (Becker et al., 2015). (Emera and Samara, 2005) proposed lumping the hydrocarbon components into volatile (C_1 & N_2), intermediate (CO_2, H_2S, C_2 to C_4) and heavy categories (C_{5+}) to calculate the minimum miscibility pressure. (Duan et al., 2013) proposed lumping several components into a single carbon number (SCN) with predetermined values for properties, such as critical pressure, temperature, and eccentric factor. Others have proposed optimal lumping schemes based on the objectives (Lin et al., 2008). In lumped model, $C_{1+}N_2$, C_2, CO_2, and H_2S, the intermediate (C_3 to C_6), heptanes plus (C_{7+}) and its properties, and temperature are the input variable. The correlation matrixes of the lumped components are provided in Figure 3. Even though the non-hydrocarbon has the least effect their contribution is important, especially the H_2S in the case of sour crudes. Both models presented by AlQuraishi (2009) and Bandyopadhyay and Sharma (2011) have eliminated the effect of H_2S.

The proposed model has the following form:

$$Ps= 82.115*(C_1+N2)-11.635*C_2+39.158*CO_2+ 38.244*H_2S -1.217*(C_3—C_6) - 890.701*(C_{7+}/MW_{7+}*\gamma_{7+}) + 4.217 *T-1042 \quad (5)$$

Where Ps is the saturation pressure in psia; and T is the temperature in °F.

Result and discussion

Model accuracy

The accuracy of the two proposed models used to calculate the saturation pressures of the previously described crude oil samples is shown in Table 3. This table shows reports the average relative error, average absolute relative error standard deviation and the correlation coefficients for all the methods considered in this study. Table 3 also reports the accuracy of the PR-EOS, SRK-EOS, Elsharkawy's model, AlQuraishi's model, and Bandoypandyay and Sharma's model and Jarrahian model. The methods proposed by Gholami et al., Ahmadi et al., and Jarrahian et al are included in Table 3, even though they produce a unreasonable estimate of the saturation pressures, as explained in the appendix.

This table indicates that the extended and the lumped models have an (AARE) in the order of 8% and 8.5%, respectively. It should be noted that the complications in the Bandoypandyay and Sharma model did not improve the accuracy of the saturation pressure, because it had a large error. It is also noted that Jarrahian et al model produces an average error of 35%. Compared with the EOS, the lumped model has a few constants and eliminates the splitting, characterizing the pseudo-components and binary interaction parameters necessary for the EOS calculations. The lumped model as well as the extended model are very simple and require one-step calculations compared with all the other methods reported in Table 3.

A cross-plot of the calculated saturation pressure using the extended model based on the training data versus the measured data is illustrated in Figure 4A. This figure indicates that most of the data falls on the unit slope line, and there is no over-fit. This trend is supported by the high correlation of regression of the model (97.6%) presented in Table 3. Figure 4B shows a cross-plot of the testing data (28 from literature and 7 new measured in PFRC at Kuwait University). The extended model indicates an excellent agreement with all of the 28 data sets chosen at random, with the exception of a few data points at higher pressures. Additionally, the model has a perfect match for the 7 newly measured samples.

Figure 5A shows the performance of the lumped training-model for capturing the measured saturation pressure for most of the 179 samples. A cross-plot of the calculated saturation pressure using the lumped model based on the training data versus measured data is depicted in Figure 5B. This figure indicates that most of the data scatters falls to the unit slope line. The lumped model also has a high coefficient of regression of 97.4%, which is indicated in Table 3.

Models Validity

The accuracy and validity of the models for estimating the saturation pressure of various types of crude oil are reported in Table 4. This table illustrates the capability of the newly proposed models to predict the saturation pressure for volatile oils, black oils, and heavy oils. Table 4; provides the compositions, properties of the heptane plus fraction, temperature, and measured saturation pressures of the samples. Additionally, the calculated saturation pressure using the PR-EOS, SRK-EOS, Elsharkawy's model (2003), AlQuraishi's model (2009), and Bandoypandhyay and Sharma's model (2011), Jarrahian et al (2015); as well as the extended and lumped models are provided in Table 4. It is clear from this table that the proposed models have a wide range of applications regardless of crude oil types. We have also included information regarding the 7 newly measured data; at the PFRC, compositional, and saturation pressure. At the bottom of Table 4, we have provided the unpublished data of the compositions; and the saturation pressures at low temperatures. It is well known that the saturation pressure of a crude sample is measured at low temperatures in addition to the reservoir temperature as a method to check the quality of the sample.

Figure 6A reports the error of all the methods considered in this paper to predict the saturation pressure for volatile oils (heptane plus content below 30%). Both the extended and the lumped models have an AARE of nearly 4% followed by Bandoypandhyay and Sharma's model (2011); and Elsharkawy's model (2003); the SRK-EOS has nearly twice the AARE of the newly proposed model. AlQuraishi's model (2009) has an AARE of 12% because the model is based on methane and neglects the effect of heavy and intermediate components, whose contribution is highly significant in the reservoir and production performance of volatile crudes. Jarrhaine et al (2015) model is not suitable for volatile oils. It shows a 35% error in estimated saturation pressure.

The performance of the extended and lumped models compared to the various methods for the black oil samples ($35 < C_{7+} < 55\%$) is provided in Figure 6B. The newly proposed models have an AARE close to 10%; followed by Bandoypandyay and Sharma's model 14%, the SRK-EOS and PR-EOS 16%, and Elsharkawy 17%. AlQuraishi's model has an AARE of 24%. Jarrhain et al (2015) model shows a 33% error in estimated saturation pressure even though the authors claim that the model is suitable for black oils.

Figure 6C shows the error level of all the methods as well as the proposed models for the newly measured crude oil samples at Kuwait University using the PFRC. Again, Jarrhain et al (2015) model show a 40% error in estimated saturation pressure.

The performance of all published models at low temperature is illustrated in Figure 6D. The low-temperature calculation indicates that Bandoypandhyay and Sharma's model (2011) has accuracy problems and is not valid at low temperatures because the authors proposed a temperature-composition related coefficient. Also, Jarrhain et al (2015) model is not suitable at low temperature. It shows a 40% error in estimated saturation pressure.

The ability of the proposed models to capture the thermodynamic is thoroughly checked against its independent variables. The validity of the model is compared to the experimentally measured data as well as the predictions by other methods used to capture the physics of increasing the saturation pressure as a function of temperature is illustrated in Figure 7. Two main points are observed here; first, the model matches the measured saturation pressures much better than the other methods. Second, it follows the physical trend predicted by the EOS's. Figure 8 shows the model performance for capturing the effect of changing the saturation pressure as a function of the molecular weight of the heptane plus fraction compared with the SRK-EOS and PR-EOS. The extended model follows the trend of the PR-EOS in increasing the saturation pressure. Elsharkawy's model (2003) was unable to predict the correct trend as predicted by the EOS's. The model sensitivity in capturing the change in the saturation pressure as a function of increasing the methane content is shown in Figure 9. The proposed model as well predicts the correct trend of increasing the saturation pressure when increasing the methane content.

Conclusion

In this paper, two models are presented to estimate the saturation pressure of a variety of crude oils when measures are unreliable or samples are unavailable. The two models are developed using compositional data and saturation pressure measurements from a large data bank comprising literature data, unpublished data, and newly measured and reported data of 231 crude oil samples. The input parameters in the first model are the extended hydrocarbon and non-hydrocarbon molar percentages and reservoir temperatures. The second model uses lumped hydrocarbon fractions of intermediate, and heptane plus fractions in addition to its properties; and the reservoir temperature as input parameters. The accuracy of the two models has been tested and compared to that of the SRK-EOS and the PR-EOS; as well as published correlations that use compositional data as the input parameters. The comparison indicates that the two models are much more accurate than all the published methods. The validity of both models has also been tested using various types of crude oils samples and compared to experimentally measured saturation pressures and estimates using the EOS as well as the published correlations. The proposed models are physically correct and much simpler than all the published methods. The models do not require tedious computations and eliminate splitting of the plus fraction, characterization of the pseudo-components; as well as the binary interaction parameters that are required for the EOS calculations.

Acknowledgement

The authors thank the General Facility Lab at Kuwait University for their cooperation and conducting the compositional and saturation pressure measurements of the 7 new samples described in this paper using Petroleum Fluid Research Center (PFRC) project no. GE 01107.

References

Agarwal, R.K., Li, Y.K., Nghiem, L., 1990. A regression technique with dynamic parameter selection for phase – behavior matching. SPE Res. Eng. February, 115 – 120.

Ahmadi, M. A., Zendehboudi, S, James, L., Elkamel, A., Dusseault, M., Chatsiz, I., Lohi, A., 2014, New Tools to determine Bubble Point Pressure of Crude Oils: Experimental and Modeling Study, *Journal of Petroleum Science and Engineering, v.123, 204-216.*

Ahmed, T., 2011, Splitting and lumping schemes of the plus fraction, working guide to vapour-liquid phase equilibria calculations, Chapter 9, Hydrocarbon Phase Behaviour, Gulf Publishing Co. 111-126.

Al-Gathe, A., Abd-el Fattah, Kh., El-Mtwally, K., 2014, New artificial intelligent approach for bubble point pressure, Int. J innovation and Sci. Research, Vol.4, No.2, July, 121-135.

Al-Marhoun, M.A. 1985. Pressure-Volume-Temperature Correlations for Saudi Crude Oils. Presented at the SPE Middle East Oil Technical Conference and Exhibition, Bahrain, 11–14 March. SPE-13718.

Almehaideb, R.A. 1992. Improved PVT Correlations for UAE Crude Oils. Presented at the Middle East Oil Show and Conference, Bahrain, 15-18 March. SPE-37691.

Al-Najjar, H.S., Al-Soof, N.B.A., and Al-Khalisy, K.M. 1988. Correlations for Bubble-Point Pressures, Gas Oil Ratios and Formation Volume Factors for Iraqi Crude Oils. *Journal of Petroleum Research* (June 1988): 13.

AlQuraishi, A. A., 2009, Determination of crude oil saturation pressure using linear genetic programing, Energy & Fuel, 23, 884-887.

Al-Shammasi, A.A. 2001. A Review of Bubblepoint Pressure and Oil Formation Volume Factor Correlations. *SPE Res. Eval. & Eng.* 4 (2): 146-160. SPE71302.

Asgarpour, S., McLauchlin, L., Wong, D. et al. 1989. Pressure-Volume-Temperature Correlations for Western Canadian Gases and Oils. *J Can Pet Tech.* 28 (4): 103.

Asoodeh, M., Bagheripour, P.,2012, Estimation of bubble point pressure from *PVT* data using a power-law committee with intelligent systems, *J. Pet. Sci. and Eng., Vol. 90–91, July 2012, Pages 1-11.*

Bandyopadhyay, P., Sharma, A., 2011, Development of a new semi analytical model for prediction of bubble point pressure of crude oils, *Journal of Petroleum Science and Engineering, Volume 78, Issues 3–4, September 2011, Pages 719-731.*

Becker, P. J., Celse, B., Guillaume, D., Dulot, H., Costa, V., 2015, Hydrotreatment modelling for variety of VGD feedstocks: a continuous lumping approach, Fuel, 139, 133-143.

Coats, K.H., and Smart, G.T., 1986, Application of regression base EOS PVT Program to laboratory data, SPERE, May 277-299.

Dansh, A. Xu, D. H., Todd, A. C., 1992 A grouping method to optimize reservoir description for compositional simulation of gas injection process, SPERE, Aug. 343-348.

De Ghetto, G. and Villa, M. 1994. Reliability Analysis on PVT Correlations. SPE-28904, Presented at the European Petroleum Conference, London, United Kingdom, 25-27 October.

Dindoruk, B. and Christman, P.G. 2001. PVT Properties and Viscosity Correlations for Gulf of Mexico Oils. Presented at the SPE Annual Technical Conference and Exhibition, New Orleans, 30 September-3 October. SPE-71633

Dokla, M.E. and Osman, M.E. 1992. Correlation of PVT Properties for UAE (United Arab Emirates) Crudes. *SPE Form Eval.* (March 1992): 41.

Droham, J.K., 1988, Enhancing the evaluation of PVT data, OSEA paper presented the 7th Offshore South East Asia, Feb. 2-5, Singapore.

Duan, J., Wang, W., Liu, H., Cong, J., 2013, modelling the characterization of the plus fraction by using continuous distribution function, Fluid Phase Equilibria, 345, 1-10.

Elam, F.M. 1957. *Prediction of Bubble Point Pressures and Formation Volume Factors from Field Data.* MS thesis, University of Texas at Austin, Austin, Texas.

Elmabrouk, S., Zekri, A., Shirif E., 2014, The prediction of bubble point pressure and bubble point oil formation volume factor in absence of PVT analysis, PET. Sci. & Tech., 32, 1168-1174.

Elsharkawy, A. M., 2003, An empirical model for estimating the saturation pressures of crude oils, Journal of Petroleum Science and Engineering 38 (2003) 55-77.

Elsharkawy, A.M. and Alikhan, A.A. 1997. Correlations for predicting solution gas/oil ratio, oil formation volume factor, and undersaturated oil compressibility. *J. Pet. Sci. Eng.* 17 (3–4): 291-302.

Emera, M., Samara, H., 2005, Use of genetic algorithm to estimate CO2-oil minimum miscibility pressure- a key parameter in design of CO2 miscible flood, J. Pet. Sci. Eng., Vol. 46, issue 1-2, 37-52.

Farasat, A., Shokrollahi, A., Arabloo, M., Gharagheizi, F., Mohammadi, A., 2013, toward an intelligent approach for determination of saturation pressure of crude oil, Fuel Processing Technology 115 (2013) 201-214.

Frashad, F., LeBlanc, J.L., Garber, J.D. et al. 1996. Empirical PVT Correlations for Colombian Crude Oils. SPE-36105 Presented at the SPE Latin American and Caribbean Petroleum Engineering Conference, Port of Spain, Trinidad and Tobago, 23–26 April. -MS.

Gharbi, R.B.C., Elsharkawy, A.M., 1999, neural network model for estimating the PVT properties of Meddle East crude oils, SPE Reservoir Eval. Eng. 2 (3), 255-265.

Gholami, A., Asoodeh, M., Bagheripour, P., 2014 How committee machine with SVR and ACE estimates bubble point pressure of crudes, *Fluid Phase Equilibria, Volume 382, 25 November 2014, Pages 139-149.*

Glasø, Ø. 1980, Generalized Pressure-Volume-Temperature Correlations. *J Pet Tech.* 32 (5): 785-795.
Heidaryan, E., Moghadasi, J., 2011, Crude saturation pressure through imperial correlation, Thermodynamics 2011, Athens, 31 August- 3 September 2011.
Hasan, M., 1993, PVT Correlations for Indonesian Crude Oils. *Proc.* 22nd Annual Convention of the Indonesian Petroleum Association, Indonesia, Vol.1, IPA93-2.1-070.
Jarrhian, A., Moghadsi, J., Heidaryan, E., 2015, Empirical estimating of black oil bubblepoint (saturation) pressure, J. Petroleum Science and engineering, (126), 69-77.
Jacopy R.H., Berry V.J.,1958, A method for predicting the pressure maintenance for reservoir producing volatile oil, AIME 213, 59-65.
Kartoatmodjo, T., Schmidt, Z. 1994, Large Data Bank Improves Crude Physical Property Correlations. *Oil Gas J.* 92 (27): 51–55.
Kartoatmodjo, T.R.S. and Schmidt, Z. 1991, *New Correlations for Crude Oil Physical Properties*, SPE 23556.
Khairy, M., El-Tayeb, S., Hamdallah, M., 1998. PVT Correlations Developed for Egyptian Crudes. *Oil Gas J.* 96 (18): 114.
Khamekhcki, E., Behvandi, R., Rashid, F., 2011, Prediction of bubble point pressure & asphaltene onset pressure during CO2 injection using ANN & ANFIS models, Pet. Sci. & Tech. Vol. 1 No.2, 35-45.
Labedi, R.M. 1982, *PVT Correlations of the African Crudes*, PhD thesis, CSM.
Labedi, R.M. 1990, Use of Production Data to Estimate the Saturation Pressure, Solution GOR, and Chemical Composition of Reservoir Fluids, SPE-21164-MS. Presented at the SPE Latin America Petroleum Engineering Conference, Rio de Janeiro, Brazil, 14-19 October,
Lasater, J.A. 1958, Bubble Point Pressure Correlations. *J Pet. Tech.*10 (5): 65–67. SPE-957-G.
Levitan, L.L. and Murtha, M. 1999. New Correlations Estimate Pb, FVF. *Oil Gas J.* 97 (10): 70.
Lee, S., 1979, Experimental and theoretical studies on the fluid properties required for simulation of thermal process, Paper SPE 8393.
Li, Y.K., Nghiem, L.X., Siu, A., 1985. Phase behavior computations for reservoir fluids: effect of pseudo-components on the phase diagrams and simulation results. J. Can. Pet. Technol. Nov. – Dec., 29 – 36
Lin, B., Leibovici, C. F., Jorgensen, S. B., 2008, optimal component lumping: problem formulation and solution techniques, Computer & Chem. Eng., 32, 1167-2272.
Macary, S.M. and El-Batanoney, M.H. 1992, Derivation of PVT Correlations for the Gulf of Suez Crude Oils. *Proc. of the* 11th EGPC Pet. Exp. & Prod. Conference, Vol. 1, 374, Cairo, Egypt,
Moharam, H.M., 1995. Prediction of viscosity of heavy petroleum fractions and crude oil using a corresponding states method. Ind. Eng. Chem. Res. 34, 4140 – 4144.
Moradi, B., Malekzadeh, E., Amani, M., Boukadi, F.H., Kharat, R., 2101, Bubblepoint pressure imperial correlation, SPE 132756.
Obomanu, D.A. and Okpobiri, G.A. 1987, Correlating the PVT Properties of Nigerian Crudes. *J. Energy Res. Tech.* 109 (4): 214-217.
Okoduwa, I. G., Ikiensikimama, S.S., 2010, Bubble point pressure correlation for Niger Delta Crude Oils, SPE 136968.
Omar, M.I. and Todd, A.C. 1993. Development of New Modified Black Oil Correlations for Malaysian Crudes. SPE-25338 Presented at the SPE Asia Pacific Oil and Gas Conference, Singapore, 8-10 February.
Owolabi, O.O. 1984. *Reservoir Fluid Properties of Alaskan Crudes*. MS thesis, University of Alaska, Fairbanks, Alaska (May 1984).

Pedersen, K.S., Thomassen, P., Fredenslund, Aa., 1988a. Characterization of gas condensate mixtures. AIChE Spring National Meeting, March 6 – 10, New Orleans, LA.

Peng, D. y., Robinson. D. B., 1976. A new two-constant equation of state , Ind. Eng. Chem. Fund. 15, 59-63.

Petrosky, G.E. Jr. 1990. *PVT Correlations for Gulf of Mexico Crude Oils*. MS thesis, University of Southwestern Louisiana, Lafayette, Louisiana.

Petrosky, G.E. Jr. and Farshad, F. 1998. Pressure-Volume-Temperature Correlations for Gulf of Mexico Crude Oils. *SPE Res Eval & Eng.*1 (5): 416-420. SPE-51395.

Riemens, W.G., Schutle, A.M., Jong, L.N.J., 1988. Birba field PVT variations along the hydrocarbon column and confirmatory field tests. J. Pet. Technol. Jan., 83 – 88

Rosengger, L., Wu., R.1990, Integrated PVT data characterization- lessons from four case histories, J. Can. Per. Tech., 38, (13), (paper 97-05)

Shojaei, M., Bahrami, F., Barati, P., Riahi, S., 2014, Adaptive neuro-fuzzy approach for reservoir oil bubble point pressure estimation, *J. of Natural Gas Sci. and Eng., Volume 20, September 2014, Pages 214-220.*

Soave, G. 1972. Equilibrium constants from a modified Redlich-Kwong equation of state, Chem. Eng. Sci. 27, 1197-1203.

Standing, M.B. 1947. A Pressure-Volume-Temperature Correlation for Mixtures of California Oils and Gases. *API Drilling and Production Practice* (1947): 275-287.

Tale, M., Alquraishi, A., Al Mushaigeh, E. 2009, New model for crude oil saturation pressure prediction using alternation conditional expectation, Offshore Mediterranean Conf. & Exhibit. Italy 25-27, 2009.

Talebi, R., Ghiasi, M. Talebi, H., Mohammadyian, M., Zendehboudi,S., Arabloo, M., Bahadori, A., 2014, Application of soft computing approaches for modelling saturation pressure of reservoir oil, *J. Nat. Gas Sci. and Eng., Vol. 20, Sep. 2014, 8-15.*

Vazquez, M., Beggs, H. D. 1980, Correlations for Fluid Physical Property Prediction. *J Pet. Tech.* 32 (6): 968-970. Vazquez, M.E. 1976. *Correlations for Fluid Physical Property Prediction,* MS thesis, University of Tulsa.

Velarde, J., Blasingame, T.A., and McCain Jr., W.D. 1997. Correlation of Black Oil Properties at Pressures below Bubble Point Pressure - A New Approach. Presented at the Annual Technical Meeting of CIM, Calgary, Alberta, 8–11 June. PETSOC 97-93.

Vogel, J.R., Yarborough, L. 1980, the effect of nitrogen on phase behavior and phase properties of reservoir fluids, SPE 8815.

Whitson, C. H., 1980, Characterizing hydrocarbon plus fraction, SPEJ, 24 (4), 683-694.

Williams, C.A., Zana, E.N., 1980. Use of the Peng-Robinson equation of state to predict hydrocarbon phase behavior and miscibility for fluid displacement. SPE 8817, SPE/DOE on Enhanced Oil Recovery, April 20–23, Tulsa, OK.

Appendix

In this appendix, sample calculations are presented to show the deficiency of some of the published methods.

Input parameters:

N_2	CO_2	H_2S	C_1	C_2	C_3	C_4	C_5	C_6	C_{7+}	GR.C7+	MWC7+	T, °F	Ps, psia
0.65	0.02	0	45.02	12.45	8.93	6.03	3.02	1.44	22.44	0.81	184	140	3002

Ahmadi et al (2014)

Ahmadi et al. proposed an 8 steps calculation (copied from their paper) to estimate the BPP as follows:

$$BPP = A \times Mw_{C7+} - \frac{B}{C_{7+}} + C \times Vol./Inter. + D \times SG_{C7+} + \frac{E}{SG_{C7+}} + \frac{F}{Mw_{C7+}} + T \times 505.967 + \frac{2.45789 \times 10^6}{T} + \frac{G}{Vol./Inter.} \quad (5)$$

$$A = 1488.46 - SG_{C7+} \times 1087.64 - \frac{601.512}{SG_{C7+}} + 1670.58 + T \times 0.40771 + \frac{6345.73}{T} + \frac{8.72863}{Vol./Inter.} \quad (6)$$

$$B = \frac{4132.53}{Vol./Inter.} + \frac{6.1977 \times 10^7}{Mw_{C7+}} + T \times 3181.39 + \frac{1.06393 \times 10^8}{T} \quad (7)$$

$$C = \frac{20964.7}{C_{7+}} - \frac{1.76162 \times 10^6}{Mw_{C7+}} - T \times 19.1963 - Mw_{C7+} \times 21.5427 + \frac{12851.5}{SG_{C7+}} - \frac{712613}{T} - 3477.49 + SG_{C7+} \times 9532.73 \quad (8)$$

$$D = 1.05345 \times 10^{+06} - \frac{1.50748 \times 10^8}{Mw_{C7+}} - T \times 676.048 - \frac{1.20252 \times 10^7}{T} - \frac{12178.2}{Vol./Inter.} \quad (9)$$

$$E = 641357 - \frac{1.02891 \times 10^8}{Mw_{C7+}} - T \times 301.034 - \frac{1.91446 \times 10^6}{T} - \frac{3873.62}{Vol./Inter.} \quad (10)$$

$$F = 2.51471 \times 10^{+08} + T \times 25332.5 + \frac{3.47893 \times 10^8}{T} - \frac{244573}{Vol./Inter.} \quad (11)$$

$$G = 20125.6 - T \times 23.9271 - \frac{703193}{T} \quad (12)$$

Ahmadi et al. define the Vol. /Inter. as the mole percentage ratio of the volatile (C_1& N_2) to the intermediate components (C_2 to C_6, CO_2 and H_2S). Application of the above equations from their paper using the input parameters results in the following:

Vol./Inter.	A	B	C	D	E	F	G	BPP
1.43211	1643.9	1545062	-271.127	45122.71	23642.47	2.57E+08	11753	1,794,123

The calculated saturation pressure (BBP) from Eq.5 is 1,794,123 psia which is unreasonable. This indicates a deficiency in the model.

Gholami et al (2014)

Gholami et al. presented 3 steps calculation (copied from their paper) to estimate the BPP as follows:

$$Tr(M) = \sum_{i=0}^{5} B_i M^i \quad (15)$$

Application of Eq. 15 resulted in the following

N_2	0.00584
CO_2	0.060917
H_2S	0.050636
C_1	-0.47658
C_2	-0.41547
C_3	-0.45083
C_4	-0.02003
C_5	0.068769
C_6	0.273921
C_{7+}	1.814523
SG.C7+	-0.0197
MWC7+	37.03525
T., F	-0.11139

Application of Eq. 16

$$Tr(P_b) = Tr(N_2) + Tr(CO_2) + Tr(H_2S) + Tr(C_1) + Tr(C_2) + Tr(C_3) + Tr(C_4) + Tr(C_5) + Tr(C_6) + Tr(C_{7+}) + Tr(SG_{C7+}) + Tr(MW_{C7+}) + Tr(T) \quad (16)$$

Resulted that Tr(Pb) = 37.8159

$$P_b = (57.0840387741 \times (Tr(P_b))^2) + (1166.9102304199 \times (Tr(P_b))) + 2207.3573204358 \quad (17)$$

Application of Eq. 17 resulted in saturation pressure (P_b) = 127,967 psia, which is unreasonable, compared to the measured one is 3002 psia. This indicates that there is a deficiency in the model.

Jarrahian et al (2015)

Jarrahian et al. proposed the following Eq. (copied from their paper) to estimate the saturation pressure:

$$P_{sat} = A_0 \exp\left(A_1 z_{N_2} + A_2 z_{CO_2} + A_3 z_{H_2S} + A_4 z_{C_1}^{A_5} + A_6 \sum_2^6 z_{C_i} + A_7 z_{C_{7+}}^{A_8} \gamma_{C_{7+}}^{A_9} Mw_{C_{7+}}^{A_{10}} + A_{11}(T - 459.67)\right) \quad (2)$$

Application of Eq. 2 resulted in saturation pressure (Ps) = 1812 psia, compared to the measured one is 3002 psia. This indicates that model has inaccuracy problem. This has been the pattern in all the crude oil samples shown in Table 4 and figures 6A through 6D.

Table 1- Summary of attempts to calculate the saturation pressure of crude oils

Method based on production information	Method based on compositional data
Standing, 1947;	Elsharkawy , 2003
Elam, 1957;	Tale et al., 2009
Lasater, 1958;	AlQuraishi, 2009
Vasquez, 1976;	Bandyopadhyay and Sharma, 2011
Vasquez and Beggs, 1980;	Khamehchi et al., 2011
GlasØ, 1980;	Farasat et al., 2013
Labedi, 1982&1992;	Ahmadi et al., 2014
Owolabi, 1984;	Gholami et al. 2014
Al-Marhoun, 1985;	Jarrahian et al 2015
Obomanu and Okpobiri, 1987;	
Asgarpour et al., 1989;	
Al-Najar et al., 1988;	
Dokla and Osman, 1990;	
Petrosky, 1990;	
Kartoatmodjo and Schmidt, 1991&1994;	
Macary and El-Batanoney, 1992;	
Omar and Todd, 1993;	
Hasan, 1993;	
De Ghetto and Villa, 1994;	
Farashad et al., 1996;	
Almehaidaeb, 1997;	
Elsharkawy, 1997;	
Velarde et al., 1997;	
Kairy et al., 1998;	
Petrosky and Farshad, 1988;	
Al-shammasi, 2001;	
Gharbi & Elsharkawy 1999	
Levitan and Murtha, 1999;	
Dindrouk and Christman, 2001	
Moradi et al., 2010	
Okoduwa and Ikiensikimama 2010	
Asoodeh and Bagheripour, 2012	
Shojaei et al., 2014	
Talebi et al., 2014	
Elmabrouk et al., 2014	
Kanu and Ikiensikimama, 2014	
Jarrahian et al 2015	

Table 2- Description of the data used to develop and test the extended model

Input	Training model (179 Data)			Testing (28 Data)			Testing (7 new experimental)		
	Min	Ave	Max	Min	Ave	Max	Min	Ave	Max
N_2	0.00	0.35	1.67	0.00	0.27	1.67	0.10	0.26	0.35
CO_2	0.00	1.27	11.37	0.00	1.17	11.37	0.14	0.41	0.83
H_2S	0.00	0.15	3.68	0.00	0.15	3.22	0.00	0.00	0.00
C_1	13.16	33.86	74.18	19.50	34.79	73.36	20.64	28.25	42.79
C_2	3.36	8.07	13.71	4.64	8.26	11.63	6.69	7.88	10.82
C_3	0.89	6.59	11.87	2.46	6.76	12.03	6.05	6.59	7.10
C_4	0.95	4.46	8.4	1.66	4.41	6.58	4.03	5.05	5.93
C_5	0.40	3.21	5.95	1.00	3.11	5.23	3.36	4.33	5.52
C_6	0.00	3.14	6.37	0.99	3.15	5.46	2.88	3.74	4.33
C_{7+}	10.72	38.89	57.73	11.18	37.88	51.22	27.61	43.49	51.31
$SG.C_{7+}$	0.74	0.87	0.959	0.70	0.86	0.93	0.85	0.89	0.92
MWC_{7+}	134	240	368.9	105	223	292	216	262	298
T, °F	58	156	319	58	165	261	121	157	230
Ps, psia	537	2198	5065	1025	2307	5000	1227	1779	3344

Table 3: Error analysis of the proposed models in comparison with all the investigated methods

Method	ARE	AARE	SD	R^2%
PR-EOS	6.24%	10.19%	1028	97.9
SRK-EOS	-3.12%	10.36%	1024	96.7
Elsharkawy (2003)	0.85%	8.57%	1015	96.4
Alquraishi (2009)	-6.06%	10.16%	1088	96.7
Bandyopadhyay (2011)	8.05%	12.28%	1054	96.8
Farasat (2013)	NA	NA	NA	NA
Gholami et al. (2014)	?	?	?	?
Ahmadi et al. (2014)	?	?	?	?
Jarrahian et al (2015)	35.52	35.56	983	95.2
New Extended Model	-0.37%	8.20%	1030	97.6
New Lumped Model	-0.50%	8.46%	1036	97.4

ARE = Average Relative Error

AARE = Average Absolute Relative Error

SD = Standard Deviation

R^2 = Correlation Coefficient

NA = Calculation was not possible, because the weight and bias were not provided.

? = The method produces unreasonable results as indicated in the appendix.

Table 4- Comparing model accuracy for calculations of saturation pressures for various crude oils

Volatile-Oil-Samples (C7+<30)																						
Fluid Source	index	N2	CO2	H2S	C1	C2	C3	C4	C5	C6	C7+	GR.C7+	MWC7+	T, °F	Ps, psi	SRK	PR	Elsh	Alq	Bandy	Jarr	Extd	Lump
Rosenegger	7	0.03	8.39	0	47.4	10.3	6.12	4.2	2.89	2.05	18.6	0.83	180	295	4000	3933	3879	4460	4420	4165	3317	4139	4178
Vogel-80	13	0.65	0.02	0	45	12.5	8.93	6.03	3.02	1.44	22.4	0.81	184	140	3002	2624	2548	2742	2895	2643	1812	3019	2996
Dansh-92	18	0	0	0	46.8	8.77	7.44	4.01	2.56	4.02	26.4	0.77	158	212	2941	2550	2594	3248	3617	3269	1944	3244	3374
	31	0.44	0.38	0	49.1	7.6	6.13	3.84	2.52	2	28	0.84	231	199	3739	3522	3507	3418	3951	3508	2309	3625	3644
Middle East	45	0.03	0.97	0	41.6	10.8	7.52	4.5	3	2.43	29.1	0.85	208	241	3069	3109	2979	3080	3513	3138	1962	3127	3140
Coats-86	51	0.55	1.03	0	36.5	9.93	8.85	6	3.78	3.56	30.4	0.84	200	234	2746	2537	2472	2714	2945	2643	1682	2777	2720

Black-Oil -Samples (35<C7+<55)																							
Fluid Source	index	N2	CO2	H2S	C1	C2	C3	C4	C5	C6	C7+	GR.C7+	MWC7+	T, °F	Ps, psi	SRK	PR	Elsh	Alq	Bandy	Jarr	Extd	Lump
Coats-86	73	1.64	0.08	0	28.4	7.16	10.5	8.4	3.82	4.05	36	0.84	252	131	1694	1674	1618	1615	1738	1644	1084	1713	1713
Middle East	88	0.08	1.37	0.69	36	8.67	5.94	2.97	1.5	3.1	39.7	0.9	274	168	2505	2710	2318	2247	2608	2321	1476	2441	2445
Middle East	114	0.29	0.48	0	28.4	8.29	7.38	5.06	3.42	4.41	42.3	0.88	252	133	1632	1672	1477	1544	1748	1517	984	1587	1598
Middle East	122	0	1.25	0	33.4	9.24	6.07	2.4	1.62	2.78	43.3	0.85	252	230	2110	2313	2278	2424	2737	2438	1410	2388	2412
Middle East	134	0	0.17	0	28.6	8.63	7.43	3.81	2.61	4.11	44.7	0.86	249	134	1548	1551	1429	1477	1764	1433	937	1503	1567
Middle East	146	0.44	0.83	0	27.8	7.54	6.9	4.61	3.1	3.11	45.7	0.9	272	133	1565	1799	1505	1558	1730	1528	1002	1602	1590
moharam-96	149	0.21	0.34	0	20	7.93	8	6.6	5.87	5.08	45.9	0.86	230	235	900	1353	1290	1531	1514	1225	773	1303	1294
Middle East	162	0.35	0.47	0	26.5	7.71	6.05	4.03	4.08	4.22	46.6	0.89	253	135	1500	1613	1380	1427	1631	1391	923	1478	1456
Middle East	176	0.12	0.51	0	28.8	7.91	6.46	3.17	2.07	3.25	47.7	0.88	250	134	1615	1665	1476	1574	1783	1502	981	1571	1614
Middle East	194	0.21	0.15	0	27.8	7.68	6.19	3.41	1.81	2.84	49.9	0.86	228	134	1590	1532	1377	1545	1680	1426	923	1466	1493
Rosenegger	207	0.37	0.02	0	17.3	6.11	8.23	5.47	3.76	3.96	54.8	0.85	242	156	537	938	888	914	1064	836	558	661	731

Newly-Measured Kuwait University Laboratory

Fluid Source	index	N2	CO2	H2S	C1	C2	C3	C4	C5	C6	C7+	GR.C7+	MWC7+	T, °F	Ps, psi	SRK	PR	Elsh	Alq	Bandy	Jarr	Extd	Lump
KU-1	209	0.33	0.22	0	25.6	6.87	6.39	5.61	4.68	4.33	46	0.879	222	133	1595	1360	1530	1442	1514	1394	860	1357	1337
KU-2	210	0.35	0.47	0	26.5	7.71	6.05	4.03	4.08	4.22	46.6	0.893	253	135	1500	1400	1620	1427	1631	1391	924	1478	1456
KU-3	211	0.27	0.74	0	23.1	7.2	6.45	5.93	4.66	4.16	47.5	0.906	290	138	1283	1140	1360	1284	1448	1296	800	1287	1219
KU-4	212	0.22	0.26	0	20.6	6.69	6.65	5.11	5.52	3.6	51.3	0.919	298	171	1227	1120	1300	1188	1415	1170	742	1176	1131
KU-5	213	0.19	0.14	0	27.8	8.37	6.88	5.28	4.33	3.85	43.2	0.89	278	121	1524	1300	1560	1402	1659	1409	928	1534	1489
KU-6	214	0.33	0.24	0	31.3	7.48	6.58	4.86	3.7	3.17	42.3	0.9	274	171	1978	1800	2200	1975	2242	1985	1224	2060	2026
KU-7	215	0.1	0.83	0	42.8	10.8	7.1	4.51	3.36	2.88	27.6	0.85	216	230	3344	3070	3400	3083	3568	3139	2003	3213	3200

Low-Temperature Oil Samples (59°F<T<71°F)

Fluid Source	index	N2	CO2	H2S	C1	C2	C3	C4	C5	C6	C7+	GR.C7+	MWC7+	T, °F	Ps, psi	SRK	PR	Elsh	Alq	Bandy	Jarr	Extd	Lump
Middle East	101	0	1.12	0.21	27	10.8	9.23	4.97	2.62	3.33	40.8	0.88	249	71	1247	1090	1033	1040	1235	564	808	1148	1205
Middle East	117	0.36	0.17	0	27.2	8.93	8.6	6.07	3.71	2.55	42.4	0.88	271	68	1130	1028	1245	1094	1248	766	820	1239	1228
Middle East	135	0	0.17	0	28.6	8.63	7.43	3.81	2.61	4.11	44.7	0.86	249	70	1270	1055	1255	1130	1311	783	826	1229	1297
Middle East	138	0.14	0.54	0	29.4	7.96	6.19	3.96	3.25	3.84	44.7	0.9	290	59	1265	1065	1350	1191	1301	880	894	1408	1390
Middle East	147	0.44	0.83	0	27.8	7.54	6.9	4.61	3.1	3.11	45.7	0.9	272	67	1320	1125	1400	1201	1268	921	879	1319	1312
Middle East	173	0.06	0.81	0	31.3	7.42	5.68	2.82	1.63	2.64	47.6	0.89	270	68	1460	1215	1535	1441	1463	1026	975	1548	1576
Middle East	193	0.16	0.3	0	24.7	7.51	6.92	4.76	2.65	3.43	49.6	0.88	239	71	1050	975	1156	1008	1107	701	714	958	986

Elsh: Elsharkawy (2003)
Extd: Extended model
Alq: AlQuraishi (2009)
Lump: Lumped model
Band: Bandyopadhyay and Sharma (2011)
Jarr: Jarrahian et. al (2015)

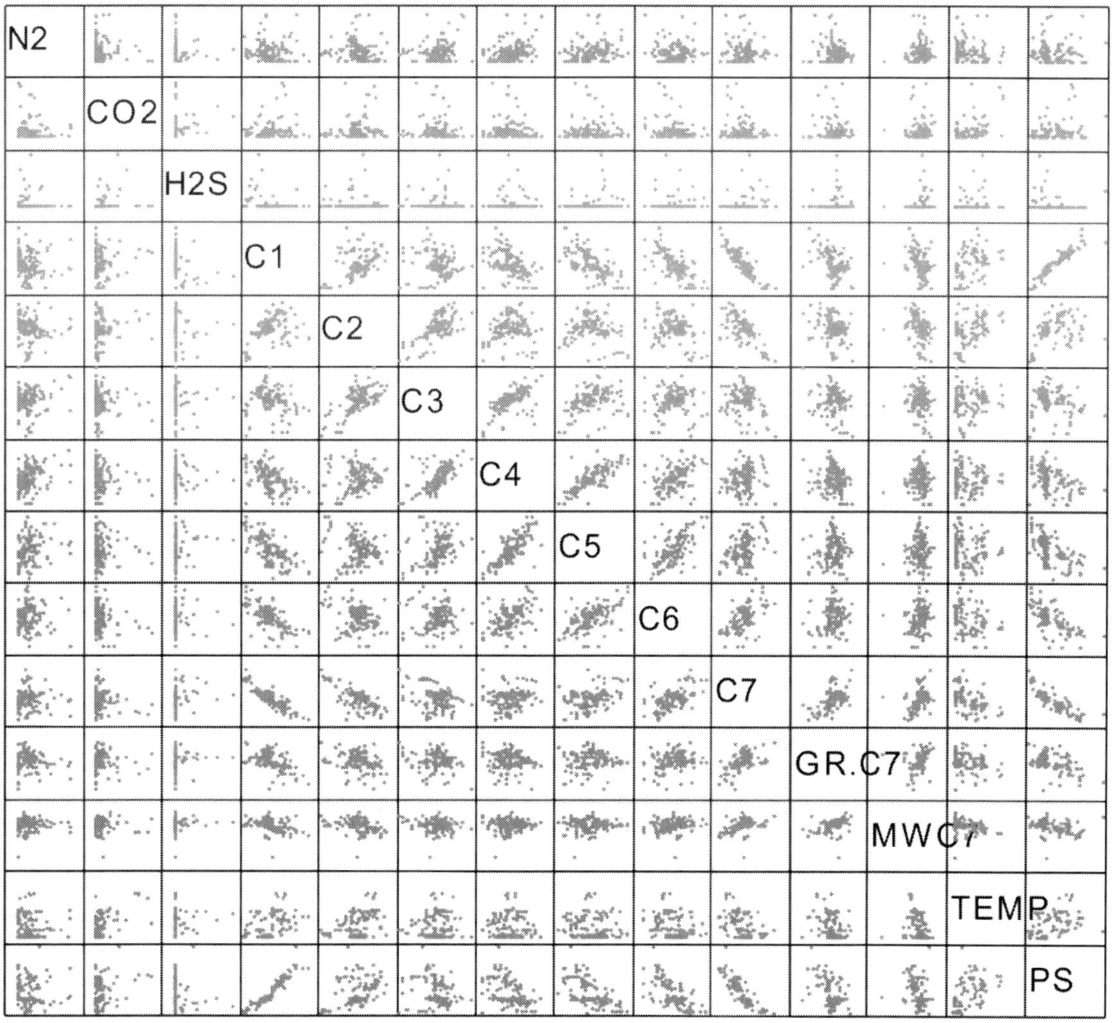

Figure 1: Correlation matrix for all the variables in the oil data bank

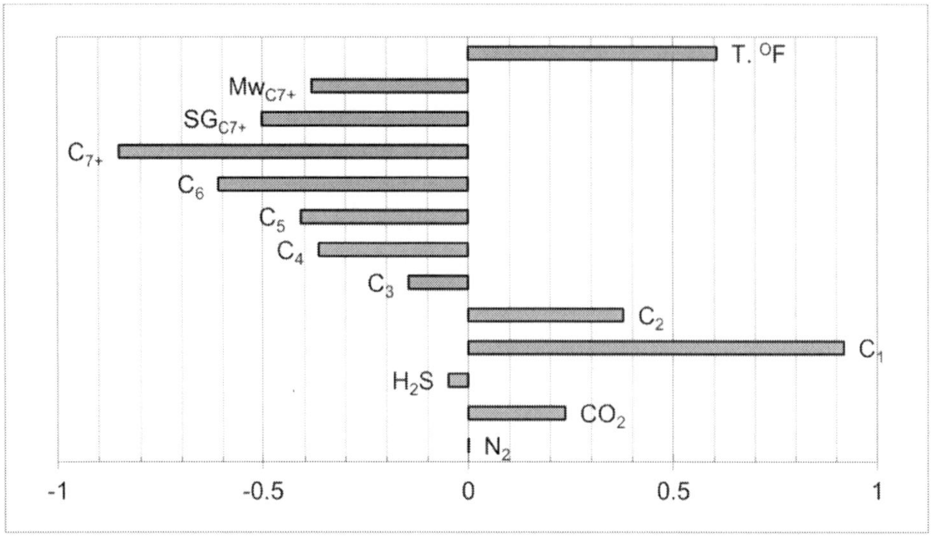

Figure 2: Correlation coefficient for all the variables in the training model (179 data)

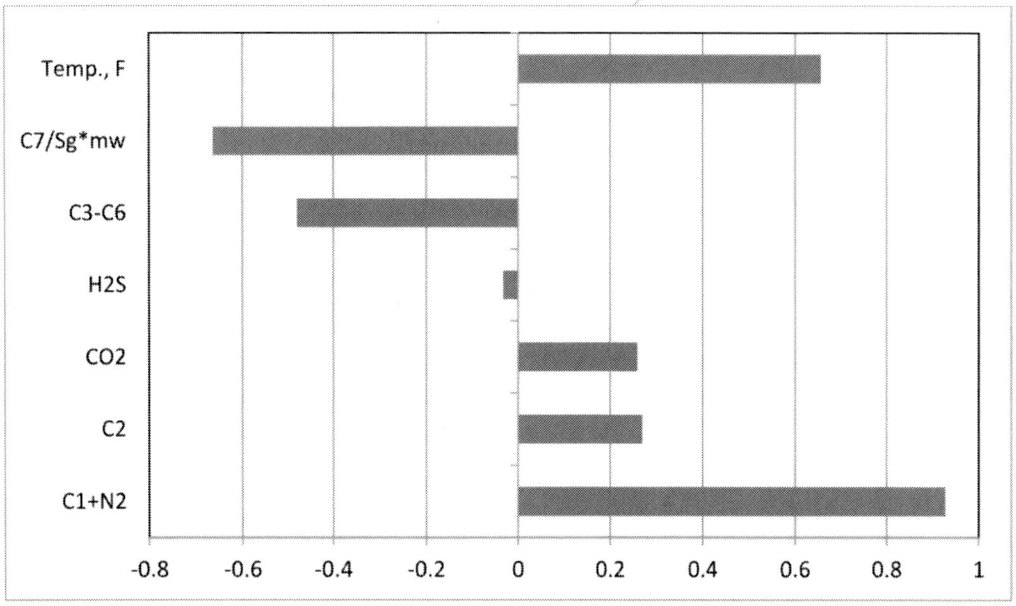

Figure 3: Correlation matrix for the lumped model (179 data)

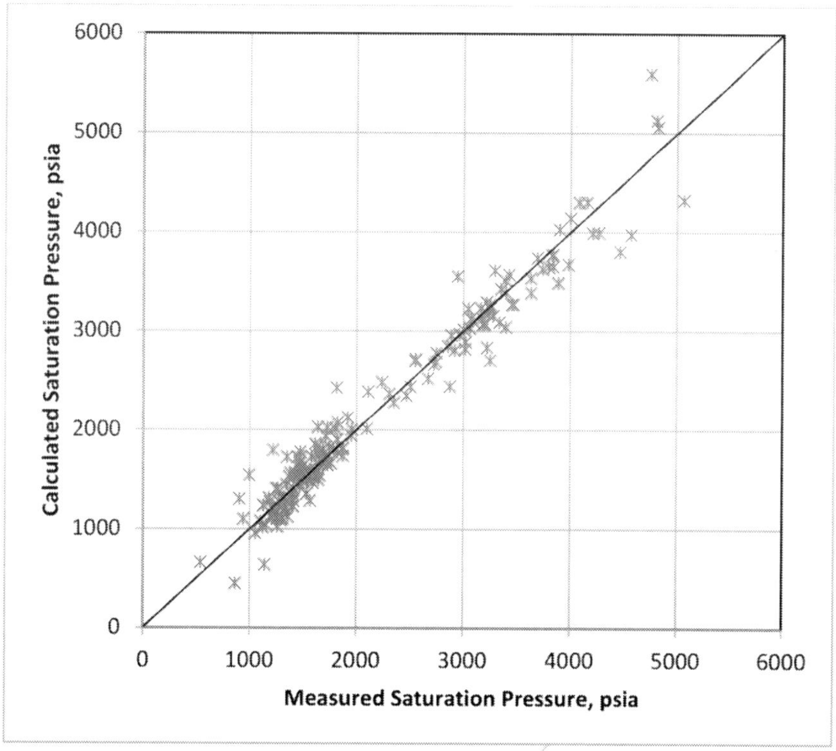

Figure 4A: Cross plot of the extended model using 179 training data

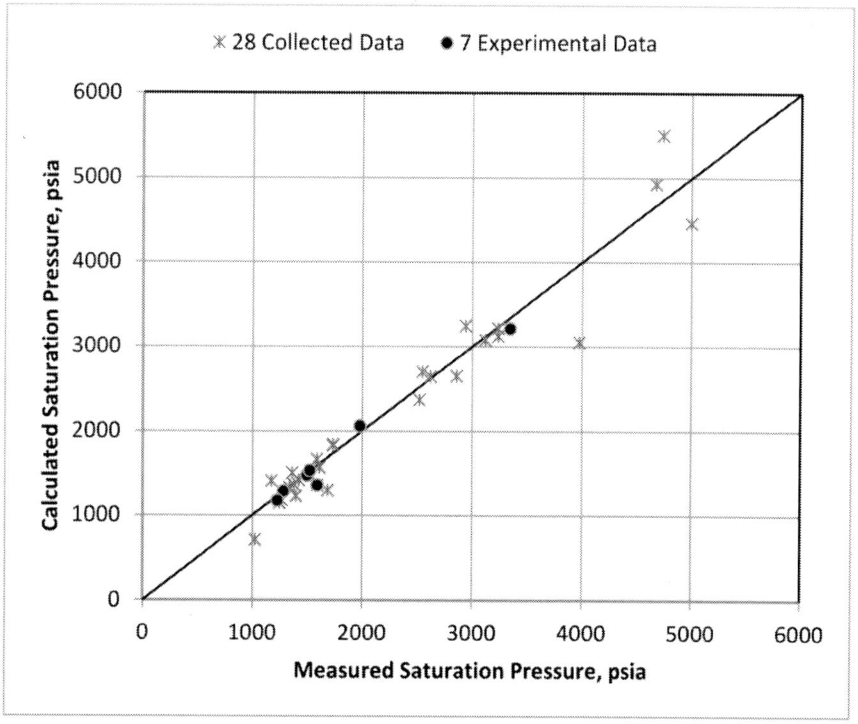

Figure 4B: Cross-plot of the extended model using 35-testing data

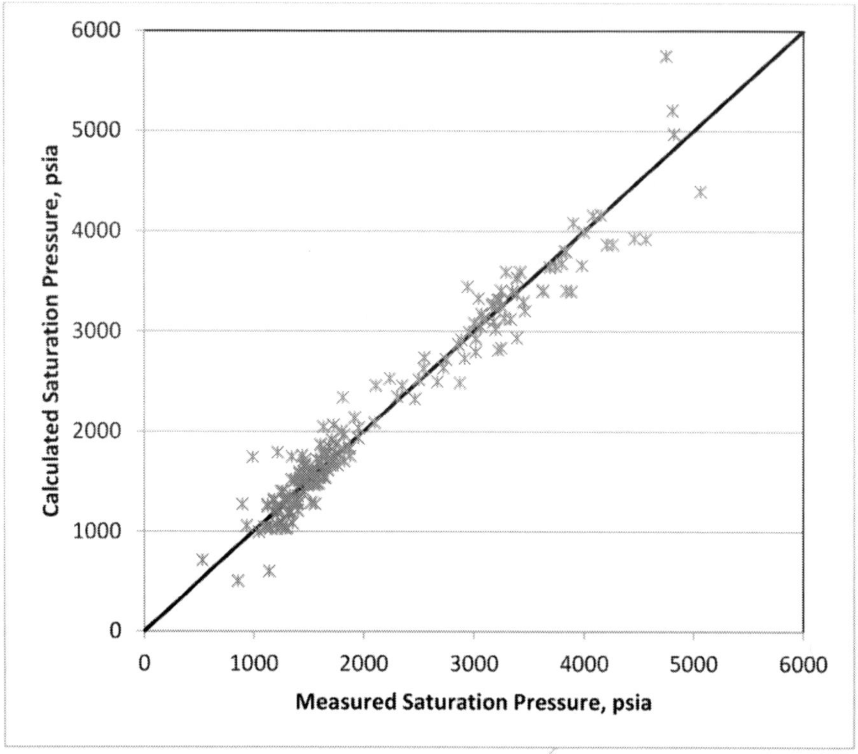

Figure 5A: Cross-plot of the lumped model using 179 training data

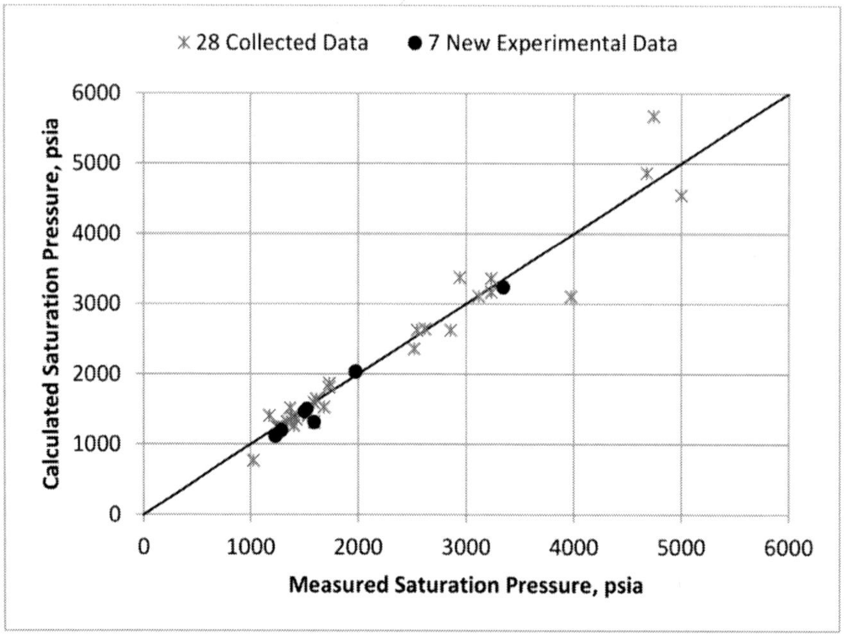

Figure 5B: Cross-plot of testing data for the lumped model

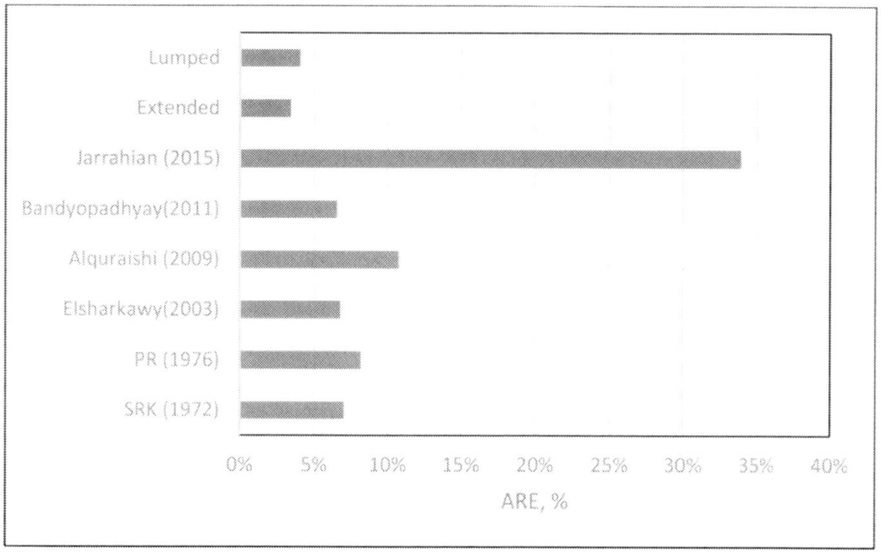

Figure 6A: Comparison of all the models' accuracy for the volatile oil samples

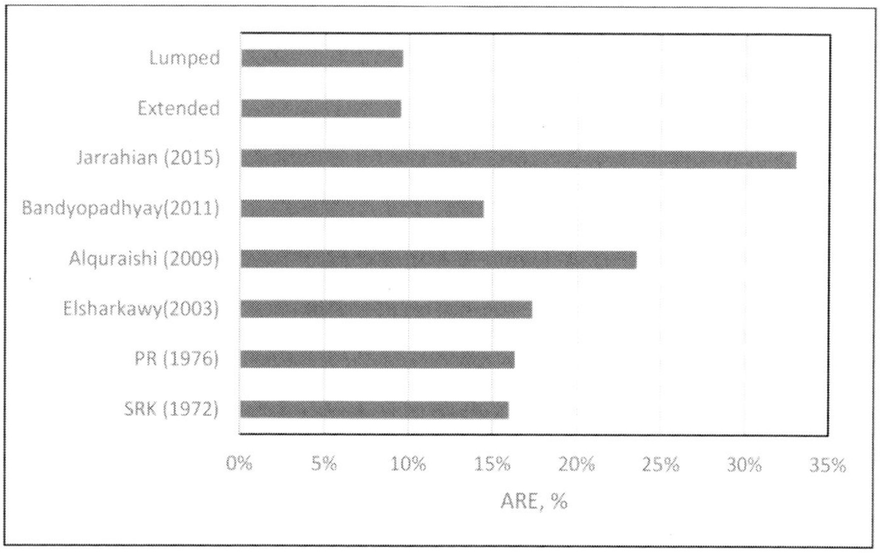

Figure 6B: Comparison of all the models' accuracy for the black oil samples

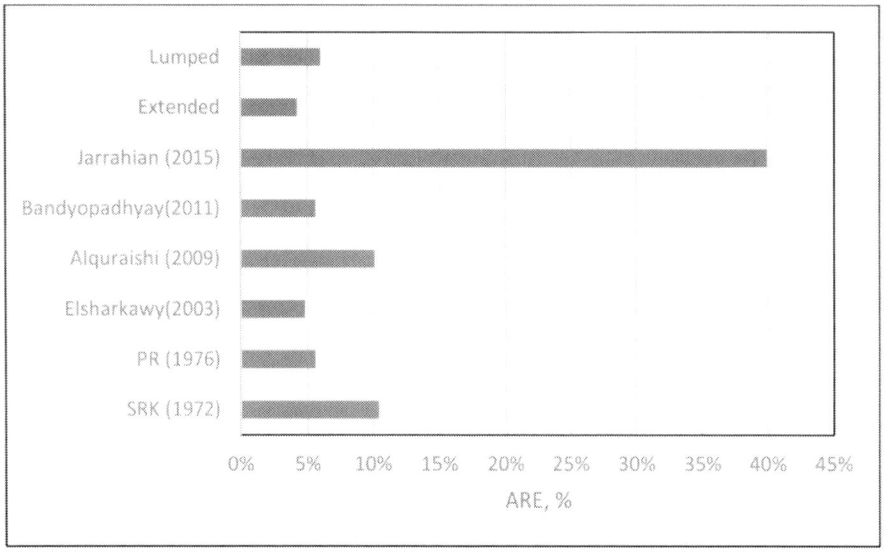

Figure 6C: Comparison of all the models' accuracy for the newly measured samples at KU

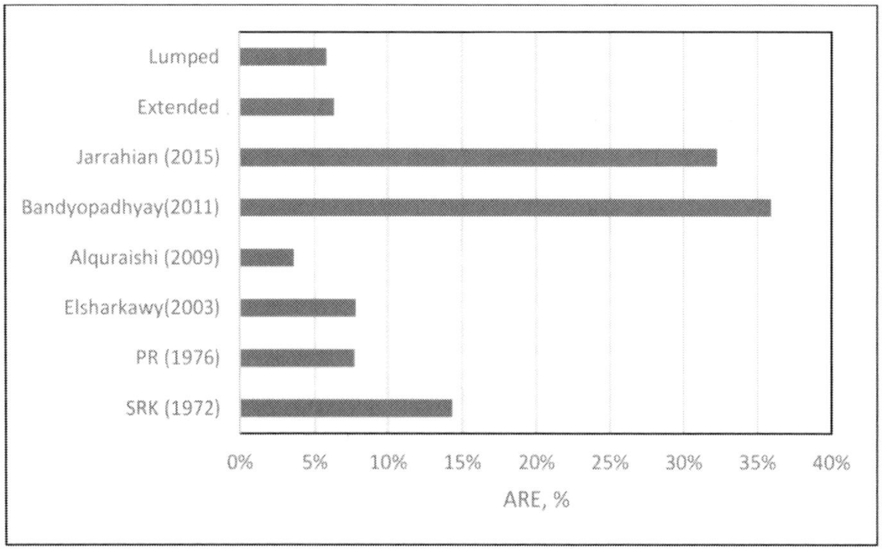

Figure 6D: Comparison of all the models' accuracy for low-temperature oil samples

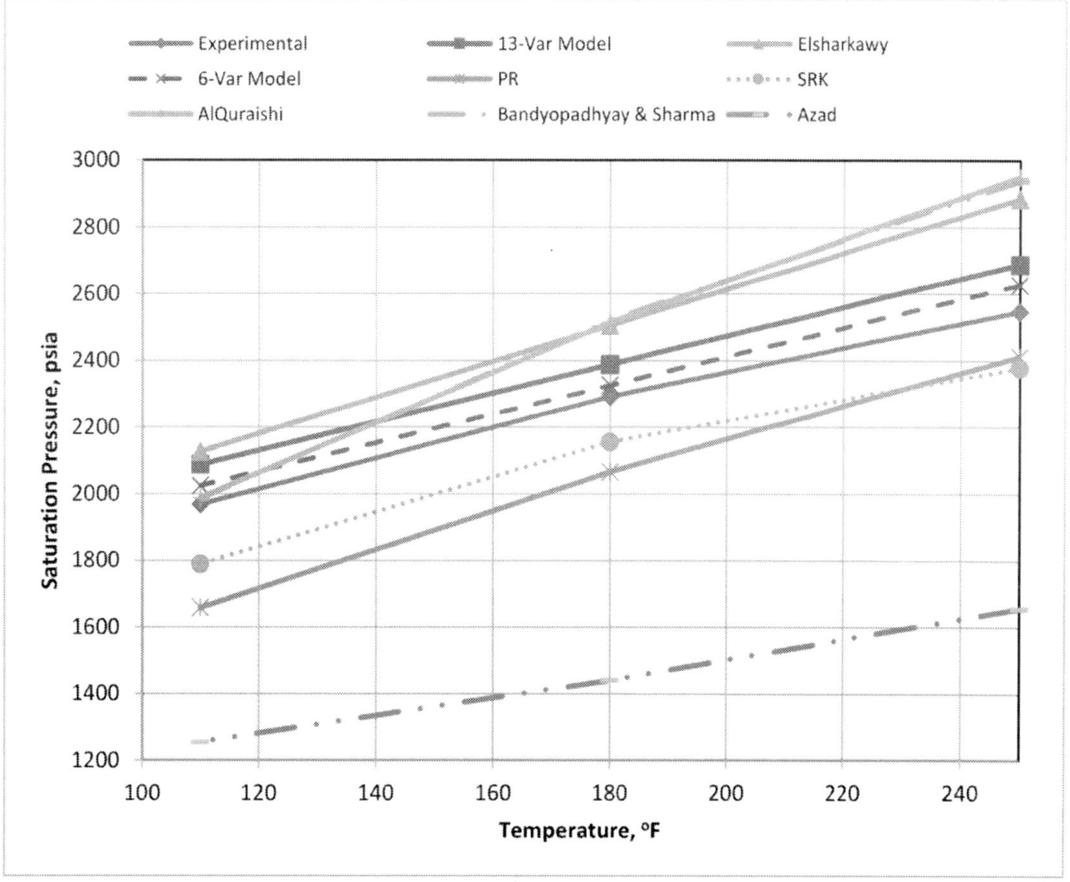

Figure 7: Changes in saturation pressure as a function of temperature for the different models (Sample #70)

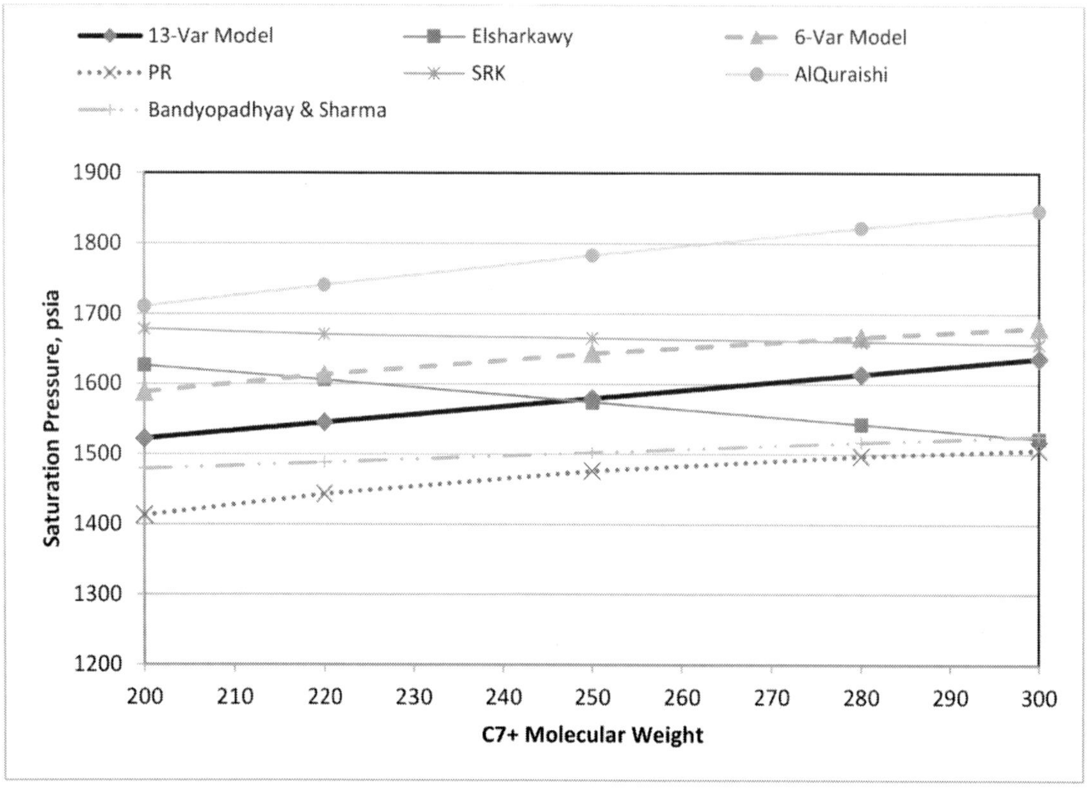

Figure 8: Comparing model performances: Changes in BPP as a function of C_{7+} molecular weight at 134°F (Sample #176)

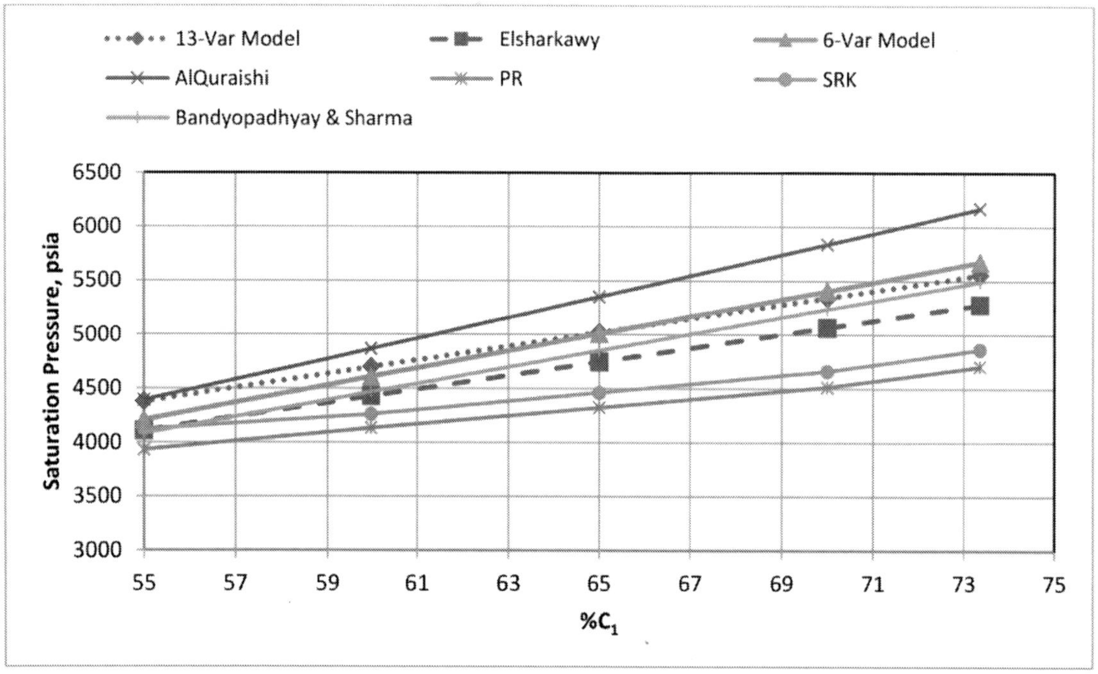

Figure 9: Sensitivity of saturation pressure to methane content at 212°F (sample #2)

439811 Purification of Ibuprofen Enantiomers By a New Technique Combining Distillation and Crystallization

Abstract:

Stripping crystallization (SC) is a new separation technique which combines distillation and crystallization. Basically, SC is operated at a triple-point condition, in which the liquid is simultaneously vaporized and crystallized due to the three-phase equilibrium. In essence, the process is continued until the liquid phase is completely eliminated and only the pure solid crystals remain in the feed. Thus, crystal washing is not required since only pure solid crystals remain in the feed and no impurities are adhered on the crystal surfaces at the end of SC. In the previous research, SC has been successfully applied to separate p-xylene from the mixed xylenes by Shiau et al. (2005, 2008, 2013).

Ibuprofen, also called 2-(isobutylphenyl)-propionic acid, is a non-steroidal anti-inflammatory drug widely used to treat headaches and minor pains. Although it is known that S-ibuprofen is responsible for the anti-inflammatory effects while R-ibuprofen might cause some long-term side effects, ibuprofen is still sold as a racemic mixture in the market due to the complexity involved in the separation of enantiomers. A thermodynamic model is developed in this work to simulate the three-phase equilibrium conditions during the SC operation. The experiments show that seeding with ultrasound mixing or magnetic stirring can be efficiently employed to purify S-ibuprofen from ibuprofen enantiomers. The experimental results, including the final enantiomeric purity and recovery ratio of S-ibuprofen, are consistent with the simulation results predicted by the model.

References:

1. Shiau, L. D.*, Wen, C. C. and Lin, B. S. "Separation and purification of p-xylene from the mixture of m-xylene and p-xylene by distillative freezing", *Ind. Eng. Chem. Res*, **44(7)**, p2258-2265, 2005.

2. Shiau, L. D.*, Wen, C. C. and Lin, B. S. "Separation of p-xylene from the multicomponent xylene system by stripping crystallization", *AIChE Journal*, **54(1)**, p337-342, 2008.

3. Shiau, L. D.*; Liu, K. F. "Investigations into the effects of the cooling rate on stripping crystallization", *Ind. Eng. Chem. Res*, **52**, p1716-1722, 2013.

Copyright © American Institute of Chemical Engineers. All rights reserved.

441535 Hydrogen Disposal Flare System Hazard Identification and Mitigation Analysis

Abstract:

A hazard assessment on a hydrogen disposal flare system was conducted in preparation for a hydrogen turbopump test at the Air Force Research Laboratory (AFRL), Aerospace Systems Directorate at Edwards AFB, CA. The test operations required the disposal of large quantities of liquid and gaseous hydrogen through a commercial flare system to prevent the formation of an explosive environment. As part of the assessment, the primary steps in the test operation process were evaluated for the formation of a potential hazard by looking at each associated failure mode and the resulting affect on personnel and facilities. Evaluation of each operational step was conducted in a Process Failure Modes and Effects Analysis (PFMEA) that identified the probability and severity of the identified effects and associated hazards. If the probability and severity were unacceptable, mitigating steps were applied until the overall hazard risks were reduced to acceptable levels. The recommendations of this assessment were incorporated into both the flare system design and the operational procedures. This paper outlines the results of the PFMEA which include identification of the effects/hazards, available risk-mitigating techniques, and implementation of these techniques used for the mitigation of the identified hazards. Based upon the findings significant design modification to the vender's commercial equipment was needed to comply with recent changes in the code. Most of the identified hydrogen flare system hazards/effects were categorized into four groups: internal explosions, obstructions, external explosions, and maintenance degradation. Many corrective actions were implemented to mitigate both the probability and severity of the identified hazards. In addition, this paper discusses the installation of oxygen monitoring systems on the flare system, determination of appropriate system purge requirements, installation of purge system status alarms, incorporation of a triple redundant flare ignition system with status indicators, and a number of other hazard risk mitigation techniques. This comprehensive description of the hazards and mitigation of a hydrogen disposal system is a plant is vital part to safe operation of both hydrogen commercial and aerospace facilities. <align="center">

Copyright © American Institute of Chemical Engineers. All rights reserved.

442068 Enhanced Furfural Production By an Innovative Bottom Dividing Wall Column with a Decanter Configuration

Abstract:

Enhanced Furfural Production by an Innovative Bottom Dividing Wall Column with a Decanter Configuration

Nguyen Van Duc Long, Le Cao Nhien, Le Quang Minh and Moonyong Lee[†]

School of Chemical Engineering, Yeungnam University, Gyeongsan 712-749, South Korea

Furfural, which are used to derive other industrial chemicals, has been identified as one of the major bio-based platform chemicals that can compete with petroleum-based chemicals. However, the current commercial furfural process has low yield and requires intensive energy. This paper reports the simulation, detailed design and optimization of an industrial process recovering furfural from lignocellulosic biomass, a renewable and inexpensive feedstock. The aqueous feed mixture from the acidic hydrolysis of biomass was fed to the separation process to recover furfural as the main product, along with methanol as valuable byproducts. The purification process was intensified by applying an innovative bottom dividing wall column with a decanter configuration (BDWC-D). The response surface methodology (RSM), which allows the interactions between variables to be identified and quantified, was used to optimize the BDWC-D structure. The predictions by the RSM showed satisfactory agreement with the rigorous simulation results. The heat pump technology and heat integration were then employed to improve the purification process. The results showed that the proposed sequence could save the overall energy requirements and total annual cost significantly as compared to the conventional sequence. This sequence can be employed for both grass-root and retrofit designs.

Keywords: Lignocellulosic Biomass; Furfural; Energy Efficiency Improvement; Dividing Wall Column; Response Surface Methodology

Acknowledgments

The authors wish to acknowledge the financial support from the R&D Convergence Program of NST (National Research Council of Science & Technology) of Republic of Korea and KITECH (Korea Institute of Industrial Technology) (ES150001).

Copyright © American Institute of Chemical Engineers. All rights reserved.

442557 Flow Field Simulation and Distribution Performance Analysis of a Gas Pipe Distributor in Fccu

Abstract:
A fluidized catalytic cracking (FCC) is an important petroleum refinery process, which, aided by highly selective zeolite catalysts, cracks large-molecule low-value heavy oil feedstock into small-molecule highvalue
products. When spend catalyst is fed into the regenerator for reactivating, it is fluidized by the gas distributor at the bottom of regenerator. The gas distributor is designed to uniformly distribute gas to maintain the fluidized state of the fluidized bed and to provide the reaction gas to guarantee the burning coke deposited on the spend catalyst. Therefore, the gas distribution performance of the gas distributor influences the quality of initial gas distribution directly. So the pipe distributor used in a industrial FCC unit, which is composed by main pipes and a number of branch pipes with many nozzle tubes, was investigated by means of flow field calculation. The gas distribution performance of the pipe distributor was analyzed based on the simulated flow field. The simulation result shows that there are a obvious nununiform
gas distribution from the different nozzle gas flow rate. The static pressure increases gradually along the flow direction in the nozzle tube while the axial flow rate decreases, which leads to the nozzle gas flow rate increasing gradually. Furthermore, an obvious bias flow occurs at the inlet of the nozzle tube, which influences the gas flowing into the nozzle. The pressure drops performance of nozzle and pressure varying in the branch pipe are analyzed comprehensively. Due to the nonuniform
gas distribution by the gas pipe distributor in the practical usage, the problems such as overlarge temperature difference among the fluidized bed layers in the regenerator have often happened, which will not only reduce burning coke efficiency and but also decrease activity of the regenerated catalyst. Meanwhile, the erosion of the distributor also occurs, and in consequence the function of the distributor fails, which even causes the nonscheduled shutdown of the FCC unit and leads to tremendous economic loss. The support is provided for optimizations of both structural parameters and operation parameters of the gas pipe distributor and improvement of the gas distribution performance. Furthermore, improvement suggestions were provided based on the above reasoning. The results are expected to be useful for the distributor design.

Copyright © American Institute of Chemical Engineers. All rights reserved.

442925 Identifying Gas Liquid Contactor Performances for Optimized Natural Gas Desacidification Amines Units - from Process Calculations to Experimental Strategy

Abstract:

Amine units are widely used for gas sweetening. The absorber performances of these reactive absorption processes are typically mass transfer limited. The proper selection of gas liquid contactor is thus critical to optimize the size of towers, the solvent flowrate and the duty of the reboiler. This work proposes an approach to identify the most suitable packings for a given application.

The case-study selected here to illustrate the approach deals specifically with a CO_2removal application where the sour gas is at 7 % vol, and the sweet gas specification is 2% vol. The gas flowrate is fixed at 0,18 MMSCFD with a solvent flowrate fixed at the minimum of the thermodynamic constraints (via equilibrium curves). An in-house rate-based simulator with accessible forms of mass transfer correlations, based on a double film model, is used to calculate the absorber height and diameter which are needed to reach the sweet gas specifications. A reference structured packing is compared with standard valve trays.

The calculations show that the bed height required with the first contactor is considerably higher than with the second one while the column diameter is reduced. In this case the use of standard valve trays is more attractive for cost optimization. Interestingly, because mass transfer is accelerated, the absorption rate with the structured packing is found to be thermodynamically limited due to a more severe temperature increase which reduces the driving force in the bottom section of the absorber.

The study determined that the optimum design for the investigated case is a compromise between a full trayed tower, that generally imposes a larger diameter, and a packed tower where the higher mass transfer forces result in a temperature rise in the bottom of the tower which in turn imposes a height increase and a much higher solvent flowrate. . To determine the mass transfer characteristics of a gas-liquid contactor providing this optimum, a sensitivity analysis has been conducted. Mass transfer parameters, i.e., liquid and gas side mass transfer coefficients and interfacial area, were varied by a factor 2 relative to the ones obtained with the reference structured packing. Calculations show that one should use a contactor that reduces liquid phase resistance, specifically, the mass transfer conductance $k_L a_e$ should be increased by at least 50%. This target could be reduced to +30% if one could simultaneously reduce the gas phase conductance $k_G a_e$ by 20%. This result can be explained by the link between mass and heat transfer (Chilton-Colburn analogy) whereby reducing $k_G a_e$decreases the temperature level which in turn thermodynamically increases the mass transfer driving force.

The next step will be to review the literature more extensively to make a short list of pre-selected random and structured packings that best meet the targets. Experimental tests will then be run on IFPEN facilities to measure flooding limits and mass transfer parameters in a small 150 mm diameter column. This step is required to confirm packing's behavior, considering the disparity of the values reported by different authors in the literature. The last step is to consolidate the existing correlations for the most promising packing, and mitigate differences between various authors by performing characterization tests on larger diameter test columns. The final correlations will be implemented in our in-house rate based simulator for industrial use.

The present approach can be applied to determine the required mass transfer characteristics for optimizing other gas treatment applications. However, the conclusions reached here as regards to mass transfer targets and the problematic temperature levels, are valid only to the present case -study, and are not directly applicable to other gas sweetening applications (selectivity, LNG, ...). For example, in a case where H_2S and CO_2 are treated simultaneously the mass transfer limitations could be very different.

Copyright © American Institute of Chemical Engineers. All rights reserved.

*** Formatting Issue - Best Available Paper/Graphic ***

443365 Technical Reconstruction of Double Effect Distillation for Methanol Process

Abstract:

Abstract: Methanol is one of important basis raw materials of chemical industry and also an excellent environment-friendly fuel, which make it as a significant commodity in the global market. Distillation is an essential unit operation for purified production of methanol. Product quality, production cost, environmental impact and energy consumption are largely related to distillation process.

Our research effort, under this circumstance, has been devoted since 2002 to energetically optimize 4-column methanol distillation system that is widely adopted in methanol units from coal, with reduced cycling water requirement. This patented process technology saw its industrial application in Chinese largest single unit of 1800Kt/a of Baofeng Group located at North China, with the lowest steam consumption of 0.9/1 t(steam)/t(methanol) compared with normally 25% higher for both simulative case study and on-site operation. Methanol product is in accordance with O-M-232G and the recycling water is reduced significantly from 67 t(CW)/t(methanol) to 52 t(CW)/t(methanol).

The flow diagram of this advanced double effect distillation process is showed in Fig.1. Its advantages arise for limitations found in current units: (1) heat exchange network failed to fully recover energy causing large CW flow to remove extra heat; (2) for pressured column, bubble point reflux need larger reflux ratio than the sub-cooled; (3) the heat of distillate from pressured column was normally ignored; (4) the low concentration of methanol of feed inlet to recovery column leads to higher reflux radio, responsible for unsatisfied unit energy efficiency.

Fig.1 The double effect distillation process.

(Column1.Light ends column; Column 2.Pressured column;

Column 3.Atmospheric column; Column4.Recovery column)

As a solution, a new flow diagram is built up and showed in Fig.2. (1) HEN totally reorganized (not completely show); (2) sub-cooled reflux to pressured column caused 10% reduction to the reflux ratio, economizing steam consumption; (3) top product of pressured column give heat to crude methanol before feeding light ends column; (4) change product and by-product position of recovery column to give efficient unit energy consumption.

Fig.2 The reconstruction of distillation process.

(Column1.Light ends column; Column 2.Pressured column;

Column 3.Atmospheric column; Column4.Recovery column)

It is estimated that the production of methanol will be over 60 million tons in 2014. If this reconstruction could be applied in the whole methanol industry, 1.3 million tons of standard coal would have been saved. This is equivalent to reduce 5 million tons emission of carbon dioxide, so the environmental impact can be largely decreased. Field operation shows reconstruction perfect substitution worth to spread to the whole methanol industry, energetically efficient and low environmentally impacting.

Copyright © American Institute of Chemical Engineers. All rights reserved.

443095 Experimental and Numerical Study of Transport Phenomenon of Falling Film Flow with Different Heating Methods

Abstract:

Falling film flow is a common phenomenon in chemical industry. It is widely used in chemical instruments, such as structured packed columns and falling film evaporators. An experiment was implemented to study the mass transfer phenomenon of binary-mixture (ethanol-water) falling film flows with different heating methods (uniform or concentrated heating). In order to study the detailed flow and mass transfer behaviors, a 2-D multiphase computational fluid dynamics (CFD) model based on volume of fluid (VOF) was developed. This model divides mass transfer into two parts: one is caused by the vapor-liquid equilibrium (VLE) and the other is caused by the heat flux. Results from simulation are inconsistent with the experimental data. Both experiments and simulations show that the ethanol concentration of the vapor with concentrated heating is 1.5% higher than the one with uniform heating.

Copyright © American Institute of Chemical Engineers. All rights reserved.

443101 Thermodynamic Modeling of HNO_3-H_2so_4-H_2o Ternary System with Symmetric Electrolyte NRTL Model

Abstract:

Nitric acid concentration/sulfuric acid concentration (NAC/SAC) process has been widely used for concentrating dilute aqueous nitric acid and recovering spent sulfuric acid. Nitric acid is concentrated by employing sulfuric acid to bind water and break nitric acid-water azeotrope at approximately 68 wt. % nitric acid. To support heat and mass balance calculations and process simulation for NAC/SAC processes, we develop a comprehensive thermodynamic model for nitric acid-sulfuric acid-water ternary system based on previously published thermodynamic models of nitric acid-water [Wang et al., Fluid Phase Equilibria, (2015) (DOI: 10.1016/j.fluid.2015.04.015)] binary and sulfuric acid-water [Que et al., Journal of Chemical & Engineering Data, 56 (2011) 963-9772] binary systems developed with symmetric electrolyte NRTL (eNRTL) activity coefficient model. The present work takes into account one-step hydration of nitric acid molecules, two-step hydration of sulfuric acid molecules, and further hydration of hydronium ions. With up to three temperature coefficients for each eNRTL binary interaction parameter and up to two temperature coefficients for each chemical equilibrium constant, the ternary model is validated with boiling point data from 383.15 to 471.45 K and activities at 273 K. Furthermore, contour plots of boiling points, vapor phase composition, and specific heat capacity of the ternary system are generated for engineering use.

Keywords:

Nitric acid, Sulfuric acid, Electrolyte NRTL model, Vapor-liquid equilibrium, Boiling point, Heat capacity, Nitric acid concentration/sulfuric acid concentration (NAC/SAC)

Copyright © American Institute of Chemical Engineers. All rights reserved.

443503 A Computational Analysis of Fire Scene for Safety in a Compartment

Abstract:

This article details numerical prediction of fire spread carried out to study the distribution of carbon monoxide concentration, smoke dispersion, soot concentration and temperature in laboratory compartment, using a commercial fire modeling software tool. The heat release rate curve predicted from the simulation was compared with literature values carried out in a road tunnel. The time variations of carbon monoxide transportation, smoke concentration and temperature spread in compartment room are reported to evaluate the effect on humans. The soot concentration was described and used throughout this study as an indicator of safety in the compartment room. The effects of vent dimension and ventilation capacity were analyzed by smoke detector response times. This work provides improvements derived from reinterpretation of actual fire scene considering the ability of mechanical exhaust system and ventilation in compartment room to mitigate the hazard of fire.

Copyright © American Institute of Chemical Engineers. All rights reserved.

443640 Preparation and Characterization of (*FURCRAEA spp*) as Green Support of Iron Nanostructured Catalyst

Abstract:

PREPARATION AND CHARACTERIZATION OF (*FURCRAEA spp*) AS "GREEN" SUPPORT OF IRON NANOSTRUCTURED CATALYST

Karen G. Bastidas G[1], Anamaria Barrera B[1], Cesar A. Sierra A[2], Hugo R. Zea R[3]

[1] Department of Chemical and Environmental Engineering. National University of Colombia (Bogota).

[2] Associate professor. Director of the Research Group of Macromolecules. National University of Colombia (Bogota).

[3] Associate professor. Director of the Research Group of Materials, Catalysis and Environment. National University of Colombia (Bogotá).

Abstract.

The availability of natural water resources has declined over time, becoming one of the biggest problems today. Environmental risk in Colombia's rivers has increase due to the widespread use of harmful chemicals in a variety of processes, ranging from illegal mineral extraction to industry wastewater, e.g. Medellin's river, due to its strategic location along a textile industrial corridor might be subject to dye contamination; the gold and platinum mining zones in Choco and Antioquia regions are being heavily affected by the use of mercury in the extraction processes. In 2009, 520 tons of mercury were imported by miners working in this area, amount that exceeds several times the mercury used by regulated industry in previous years (Ministry of Mines and Energy of Colombia, 2007). Not only the environmental risk is a concern but also the economic conditions of many low-income families whose livelihood in many cases depends exclusively of the of the mining.

Due to the issues raised above, the main goal of this study is to research, develop and implement innovative technologies in the synthesis of highly efficient, low cost and environmentally friendly iron based catalysts capable of degradate pollutants, ensuring operational stability. A native fiber (Fique, FURCRAEA spp) has been selected as a solid support for these catalysts due to its abundance, low cost and special physical and chemical characteristics.

The combination of cellulosic fibers with iron oxide nanoparticles could provide exceptional biodegradable composite materials for the treatment of organic dyes and heavy metals present in wastewater. Likewise, the use of iron particles immobilized on fique fibers allows an easy separation catalyst-treated water, turning the biocatalyst in highly recyclable and reusable material.

Fique fiber was initially characterized by determining the percentage of ashes, total lignin, soluble and insoluble lignin, cellulose, hemicellulose, total solids and humidity, obtaining 0.94, 22.58, 0.5, 22.8, 36.3, 27.5 0.701 and 2.1%, respectively. Additionally, the isoelectric point (pzc), acidic and basic sites were measured.

Then, the fiber was subjected to a pretreatment in order to expose more active sites which on time serve as nucleation points for the deposition of nanoparticles. This was done by means of an ultrasound-assisted procedure, varying the sonication time (30, 60 and 90 minutes). The effect of the sonication was followed by measuring the contents of lignin, cellulose and hemicellulose within the fiber after each time. 60 minutes of sonication was found to be most beneficial time since the amount of lignin and hemicellulose present in the fiber was reduced, while conversely the amount of crystalline cellulose exposed increases.

In order to improve the chemical ability of the nucleation sites to interact with iron and then to form stable and size controllable nanoparticles, a cationization procedure was performed on the fiber; this was achieved by the exposure of the fiber to acidic and basic conditions promoting cationization. Three cationization times (1, 2 and 3 hours) were studied and their effect on the fiber chemical surface was evaluated using FTIR-ATR and XRD analysis. The results showed 3 hours as the optimal cationizing time.

Subsequent to the cationization of the fiber surface, a procedure of ionic/cationic exchange was implemented by exposing the fiber to a solution of iron salts in order to replace the cations in the fiber surface and anchoring the iron nanoparticles. Different conditions were studied in order to synthetize zero valent iron nanoparticles (nZVI) and others with different oxidation states. Finally, to track the impregnation process on the fiber, samples from the precursor solution were taken at different times and its iron concentration was determined by atomic absorption spectrophotometry (AAS). Complementary characterization performed by X-ray diffraction, BET surface area and electronic microscopy, among others.

Key words: Natural support (fique), iron nanoparticles, bionanocatalyst.

Copyright © American Institute of Chemical Engineers. All rights reserved.

443764 Optimization of Heat Exchanger Network in Steam Assisted Gravity Drainage Oil Sands Surface Facilities

Abstract:

Steam assisted gravity drainage (SAGD) is the most viable and environmentally safe recovery technology for extracting heavy oil and bitumen. It is a thermal in-situ production process where parallel wells are drilled horizontally into an underground bitumen reservoir. Steam is produced at the central processing facility (CPF) and injected into the reservoir through the wells. The steam heats the bitumen to a point where gravity allows it to flow down to the lower well where the mixture of bitumen and water is then pumped to the surface. The water and bitumen are separated at the CPF. Typically, more than 90% of the energy consumed in producing a barrel of bitumen using SAGD is used to generate the steam that is injected into the bitumen reservoir. Fossil fuels is employed in the process, which causes GHG (greenhouse gas) emissions. The usage of hot and cold utility is reduced by increased heat exchange and optimization of the heat exchanger network, which results in reduction of GHG emissions. In this work, pinch analysis is performed to minimize energy consumption.

Copyright © American Institute of Chemical Engineers. All rights reserved.

443774 Thermodynamic Anaysis of Dehydration of n-Propanol Using Heterogeneous Azeotropic Distillation

Abstract:

Heterogeneous azeotropic distillation is a widely popular process in chemical process industries for the separation of binary azeotropic mixtures into their constituent pure components. It involves the addition of an entrainer which is partially miscible and forms a heterogeneous azeotrope preferably with one of the original components. This is a powerful technique which combines several physical phenomena such as presence of azeotropes and liquid-liquid immiscibilities, and is used widely in industries to enhance difficult or intractable distillation separations. For the design of heterogeneous azeotropic distillation columns, a detailed study of the heterogeneous region is crucial to plan the distillation sequence, which requires experimental vapor-liquid-liquid equilibrium (VLLE) data.

The present study focuses on the investigation of the heterogeneous region of the well known industrially available azeotropic mixture n-Propanol-Water system using Propyl acetate as entrainer. The VLE data of the system was generated by means of modified Othmer still and the still is modified using a probe type ultrasonic homogenizer to account for the heterogeneity. Isobaric vapor- liquid- equilibria and vapor - liquid - liquid - equilibria data at atmospheric pressure have been obtained for the ternary system n-Propanol-water-propyl acetate and the constituent binary systems. The equilibrium data has been correlated using NRTL, UNIQUAC and UNIFAC models and compared with the experimental data.

Copyright © American Institute of Chemical Engineers. All rights reserved.

443776 Study of Ionic Liquids As Entrainers for the Extractive Distillation of *N*-Propanol and *N*-Propyl Acetate Binary Azeotropic Mixture and the Isobaric VLE Studies of the System

Abstract:

Distillation is a widely accepted technique for the separation of binary liquid mixtures. Separation by simple distillation is not possible for binary mixtures that form azeotropes. One of the commonly used methods to separate azeotropes is extractive distillation. It is the process of adding an entrainer that can alter the relative volatility of the mixture and thereby breaking the azeotropic point. Ionic liquids on account of their physiochemical properties as well as the recyclability have been used as a preferred replacement for the conventional entrainers. Extractive distillation with ionic liquids has the advantages of both liquid solvent (easy operation) and solid salt(high separation ability). The corrosion and dissolution problems by using inorganic salts can be rectified by these non-volatile, chemically stable, high electrical conductive molten salts.

The azeotropic system under consideration in the present study is *n*-propanol - *n*-propyl acetate mixture. These compounds which constitute the products/reactants of esterification reaction of n-propanol form a minimum boiling azeotrope at 94.7°C. *n*-propyl acetate has been included in a list of volatile acetates as an ecologically benign CFC-free solvent for aerosol sprays.

Vapour liquid equilibrium(VLE) measurements of *n*-propanol and *n*-propyl acetate are performed at atmospheric pressure in a modified Othmer still which is a continuous distillation still in which vapour phase is recirculated. The most suitable ionic liquids for shifting of azeotropic composition of the system are selected using COSMO-RS model with suitable combinations of anions and cations. The isobaric VLE of the system with [EMIM][HSO_4] is obtained and compared with the existing data. The influence of alkyl chain length on the performance of ionic liquids is also analysed with regard to the previous studies. The effect of polarity is the most important factor on the behaviour of ionic liquid solvation in alcohols. Experimental VLE data were correlated using modified UNIFAC model.

Copyright © American Institute of Chemical Engineers. All rights reserved.

443784 Measurement and Correlation of the Mutual Solubility of Diisopropylamine + Water and Triethylamine + Water Systems at High Pressure

Abstract:

The mutual solubility data of diisopropylamine + water and triethylamine + water systems were measured at the pressure up to 100 MPa and the temperature range was 280.2~316.2 K. Experiments were performed in a high-pressure autoclave with a moveable piston. The experimental data were correlated satisfactorily with NRTL activity coefficient model and the binary parameters were obtained.

Copyright © American Institute of Chemical Engineers. All rights reserved.

443819 Improving High Shear Rate Friction Reduction Behavior of Polyacrylamide-Based Fracturing Fluids Using Xanthan Gum

Abstract:

The main component of slickwater in fracturing low permeability formations is a high molar mass polymer such as polyacrylamide (PAM), which acts as a drag reduction agent in turbulent flow. Polyacrylamide does not degrade in the environment and also suffers from mechanical degradation at high turbulent flow rates. Five different concentrations of PAM as well as mixtures of polyacrylamide/xanthan were prepared to investigate the possibility of improving shear stability of PAM. Drag reduction measurements were performed in a flow loop and friction factors at different flow rates were calculated using rheological parameters. Based on the observations, Xanthan (XG) is not as good a friction reducer as PAM, even at three-fold higher concentration. XG slightly improved the shear stability in mixed polymer solutions at low concentrations. But at high concentrations, XG significantly improves the friction reduction behavior of PAM at nonionic and ionic environments.

Copyright © American Institute of Chemical Engineers. All rights reserved.

443836 Oxidative Dehydrogenation of n-Butane on Carbon Catalysts

Abstract:

Butenes and butadiene, which are one of the important industrial chemical products to produce synthetic rubber , are sythesized traditionally by oxidative dehydrogenation (ODH) of n-butane over various mixed metal oxides. Metal-free carbon nanomaterials has recently received much attention by researchers as a catalyst for the ODH reaction due to its environmental acceptability, corrosion resistance, and unique surface properties. For the reason, the reaction system was modelled in this study for the sythesis of butene and butadiene. For the modelling of the reaction system, the catalytic performance of several carbon catalysts, such as graphite and activated carbon, were initially tested. Then, the reaction mechanisms and kinetics were developed based on experimental results. Lastly, the mathematical modeling of the reaction system was optimized to maximize an overall yield of the main products. This approach allowed us to develop the optimized reaction system more efficiently.

This research was respectfully supported by Engineering Development Research Center (EDRC) funded by the Ministry of Trade, Industry & Energy (MOTIE). (No.N0000990)

Copyright © American Institute of Chemical Engineers. All rights reserved.

443855 Modeling the Thermal Behaviors of an Ultracapacitor Module

Abstract:

Ultracapacitors, also known as supercapacitors, have the potential to meet the increasing power requirements of the sustainable energy systems for the grid-scale energy storage systems and electrified vehicles. As compared to batteries, ultracapacitors offer a higher power density, higher efficiency, and longer shelf and cycle life. Because the performance of an ultracapacitor depends on temperature, it is important to calculate accurately the thermal behavior of a single ultracapacitor cell and the module composed of multiple ultracapacitor cells for the efficient and reliable systems integration of an ultracapacitor in grid and transport applications.

In this work, modeling is performed to study the effect of the layout configugation of the ultracapacitor cells on the thermal behaviors of an ultracapacitor module. The ultracapacitor module is subject to the charge/discharge cycling with constant-current between the nominal voltage of the ultracapacitor module and the half of this voltage. The charge/discharge current values examined are 50, 100, 150, and 200 A. The validation of the modeling approach is provided through the comparison of the modeling results with the experimental measurements.

Copyright © American Institute of Chemical Engineers. All rights reserved.

443856 Modeling the Electrolyte Flow Behaviors of a Zinc/Bromine Flow Battery Module

Abstract:

The zinc/bromine (Zn/Br_2) flow battery is a promising rechargeable power source for energy storage systems because of its inherent chemical simplicity, high degree of reversibility at the electrodes, good energy density, and abundant low-cost materials. The Zn/Br_2 flow battery system, intended for the grid-scale energy storage system, is to be made in modular form. A Zn/Br_2 flow battery module is composed of the multiple stacks consisting of bipolar carbon electrodes. Since the stacks require electrolyte circulation on both the negative and positive sides, two plumbing loops are provided. The flow loops contain an electrolyte storage tank and a circulation pump. A heat exchanger may be included for temperature control in the flow loop. The design of the electrolyte circulation plumbing determines the distributions of electrolyte flow and pressure within the stacks and the efficiency of the stack is dependent on the flow distribution throughout the stack. The modeling of the distributions of electrolyte flow and pressure within the module is needed to optimize the plumbing for electrolyte circulation and thereby to improve the performance of the module. In this work, modeling is performed to study the effect of the layout configuration of the plumbing for electrolyte circulation on the distributions of electrolyte flow and pressure within a 50 kWh Zn/Br_2 flow battery module composed of four stacks. It is seen that the plumbing layout for electrolyte circulation can have a significant effect on the flow rate distribution of electrolyte through the stacks.

Copyright © American Institute of Chemical Engineers. All rights reserved.

443837 Blowdown Dynamic Modeling and Analysis of Pressure and Temperature

Abstract:

Vessel blowdown is one of the safety operation procedures that discharges flammable gas or liquid to flare system during emergency situation such as fire. When blowdown occurs, the liquefaction and freezing effect cause a pressurized vessel to be brittle, exposing it to possible hazard situation. Also the blowdown systems are recommended to be accordance with API(American Petroleum Institute) 521 code. API 521 requires pressure in vessel should be half of the initial pressure or be lower than 100 psig in 15 minutes after commencing blowdown in fire case. Therefore, accurate estimation over the change of temperature and pressure of the vessel during blowdown period is crucial for the design of blowdown system. For the reason, researchers have developed a few dynamic blowdown models so far. However, the model was found to be rather inaccurate under certain conditions. Therefore a new dynamic blowdown model was developed in this work. For the accurate analysis, the model algorithm considers both vapour and liquid phase in non-equilibrium condition that the temperatures of vapour and liquid phases are different. Futhermore, heat transfer from the vessel wall to discharge gas at either laminal or turbulent flow in the vessel was considered in the model to increase the accuracy of blowdown estimation. For thermodynamics, both Peng-Robinson and Soave-Redlich-Kwong equations of state were applied to the model, and the results were compared each other. Finally, the simulation results were compared with Haque et al's experimental and steady-state simulation results.

Copyright © American Institute of Chemical Engineers. All rights reserved.

443974 Observer-Based Model Predictive Control for Time-Delay Systems

Abstract:

Model predictive control is one of the most widely used advanced process control technologies in industry.[1] It has many capabilities such as dealing with input/state constrains and uncertainties. Observer-based model predictive control was developed to estimate states information when the states are not measurable directly. Many chemical processes have time delays in either states or measurements, such as polymer production and some bioprocesses. [2] In this presentation, we are proposing an unknown input observer based model predictive control algorithm for time-delay systems. Unknown input observer is an observer design to eliminate certain types of unknown inputs in the system. Stability of unknown input observer-based model predictive control without time delay was provided by our previous study. Terminal cost and terminal region are designed for a class of non-linear time delay systems. The close-loop stability of the controller is provided. [3][4]

An example of bioreactor is provided which has uncertain time delays in measurements but with a known upper bound. Two measurements are provided which are online biomass and offline ethanol measurements. Model-plant mismatch is considered as unknown inputs in the system. State estimation and control are calculated using the observer and an optimization algorithm.

Reference:

[1] Marcus Reble et al., Model predictive control of constrained non-linear time-delay systems. IMA Journal of Mathematical Control and Information, 2010, 1-19

[2] Seung Cheol Jeong and PooGyeon Park, Constrained MPC algorithm for uncertainty time-varying systems with state delay, IEEE TRANSACTIONS ON AUTOMATIC CONTROL, 50, 2005, 257-262

[3] Mayuresh Kothare et al., Robust constrained model predictive control using linear matrix inequalities. Automatica, 32, 1996, 1361-1379

[4] Wook Hyun Kwon et al., A simple receding horizon control for state delayed systems and its stability criterion, Journal of Process Control, 13, 2003, 539-551

Copyright © American Institute of Chemical Engineers. All rights reserved.

444020 Thermodynamic Modeling of Aqueous Lithium Sulfate and Its Solubility in the Presence of Na+, K+ and Mg2+ Sulfates

Abstract:

This work is an extension of the previous works done by our group towards development of a rigorous and comprehensive thermodynamic model for high salinity produced water that is essential for mass and heat balance calculations and simulation and design of novel treatment processes. Specifically this work presents a thermodynamic model that accurately represents all thermodynamic properties of aqueous lithium sulfate solution including the solubility of Li2SO4 in water and also in the presence of Na+–K+–Mg2+ sulfates. By explicit consideration of solution chemistry of each subsystem, symmetric electrolyte NRTL model is used to correlate available experimental data of all pertinent thermodynamic properties including activity coefficient, osmotic coefficient, excess enthalpy, and liquid heat capacity over the entire concentration range. This facilitates quantification of eNRTL binary molecule–molecule, molecule–electrolyte, and electrolyte–electrolyte interaction parameters. Also, solubility data are used to regress the thermodynamic constants required for modeling the liquid–solid equilibrium of lithium sulfate in water and in the presence of Na+–K+–Mg2+ sulfates. Displaying a significant prediction ability, the model accurately represents the thermodynamic behavior of the system from infinite dilution up to saturation and temperatures from 298.15 K to 573.15 K.

Copyright © American Institute of Chemical Engineers. All rights reserved.

444044 Organic Solar Cells

Abstract:

This paper describes highlights the need of next-generation "organic solar cells". There is a continuing need for organic solar cells and arrays with very high specific power densities (1000-5000 kW/kg) for generating power in a new generation of ultra light space payloads example in NASA. An emerging technology with promise to meet these ambitious goals are organic solar cells based on very thin films of discrete layers. Theoretical efficiencies for multi junction solar cells are more ~70% as compared to ~30% for single junction devices. There is continuous development going on high mobility, band gap engineered organic donors and acceptors, as well as development of three junction devices and scaled up devices on lightweight flexible polymer substrates. The main interest in this paper is organic semi-conducting crystal: Pentacene, which has photo-conducting states. Organic solar cells manufactured today have low efficiency other than their stability issues. The efficiency of organic solar cells can be improved by either designing polymer materials that absorbs more of solar spectrum or by minimizing the charge collection efficiency by controlling the morphology of polymer films. Once fully developed, organic solar cells would revolutionary power the industry due to their flexibility and minimal weight

Copyright © American Institute of Chemical Engineers. All rights reserved.

444055 A Refined Non-Random Two-Liquid Segment Activity Coefficient Model for Solubility Modeling

Abstract:

Accurate correlation and robust prediction for complex molecule solubility or solid-liquid equilibrium (SLE) in solvents and solvent mixtures have drawn much interest recently [Tanveer et al., *Chem. Eng. Prog.* **2014**, 110 (9), 37-47]. Solubility models such as the Non-Random Two Liquid Segment Activity Coefficient (NRTL-SAC) model [Chen and Song, *Ind. Eng. Chem. Res.* **2004**, 43, 8354-8362] are now routinely used in the industries. However, an important unfilled need is to model the solubility in oligomers/polymers/surfactants such as asphaltene nanoaggregates in petroleum fluids and drug molecules in lipid-based or solid dispersion formulation [Tung, *Org. Proc. Res. Dev.* **2013**, 17, 445-454]. This paper presents a refinement to the very successful NRTL-SAC model to better account for the contribution due to size differences of molecules. Specifically, the Flory-Huggins combinatorial entropy expression for conceptual segments in the model is substituted with the Staverman-Guggenheim expression for real molecules. We present the rationale for the refined model, the updated conceptual segment numbers for common solvents, and the newly identified conceptual segment numbers for common drug excipients. The model performance is illustrated by prediction of drug molecule solubility in various lipid solvents and solvent mixtures, compared to experimental data and previous work by other models. The refined NRTL-SAC model shows enhanced capability to correlate and predict drug molecule solubility in common solvents and excipients used in drug formulation.

Copyright © American Institute of Chemical Engineers. All rights reserved.

444057 Explaining the Paradox of Asphaltene Precipitation with n-Alkanes

Abstract:

Explaining the paradox of asphaltene precipitation with n-alkanes
Md Rashedul Islam and Chau-Chyun Chen
Department of Chemical Engineering, Texas Tech University

Abstract
Paraffins, specifically alkanes, promotes asphaltene precipitation when they are blended with asphaltene-containing petroleum fluids. The alkane/oil volume ratio required for onset of asphaltene precipitation increases with the carbon number and yields a maxima. Experimental study indicates that the maxima lays at C7[1] or between C8 -- C10 [2]. However, precipitation yield decreases asymptotically with increase of the carbon number even if excessively high alkane/oil volume ratio is used [3]. For example, n-pentane be a better solvent for asphaltene than n-hexane in regard of both onset of precipitation and precipitation yield. However, high carbon number alkane beyond the maxima, n-hexadecane for instance, can be a better solvent at onset of precipitation point and a poorer solvent in regard of precipitation yield than n-pentadecane. A thermodynamic explanation for this paradox is necessarily important to find the compatibility of bitumen or crude oil with paraffinic hydrocarbons. In this work, this paradox is explained by the aggregation thermodynamics of asphaltene precipitation [4]. Both the enthalpy and entropy of mixing are investigated to explain the effect of paraffin carbon number on asphaltene precipitation. Athabasca and cold lake bitumen samples are studied.

References
[1] G. Hotier and M. Robin, "Action de divers diluants sur les produits pÃ©troliers lourds: mesure, interprÃ©tation et prÃ©vision de la floculation des asphalténes," Oil & Gas Science and Technology, vol. 38, pp. 101-120, 1983.
[2] I. A. Wiehe, H. W. Yarranton, K. Akbarzadeh, P. M. Rahimi, and A. Teclemariam, "The paradox of asphaltene precipitation with normal paraffins," Energy & fuels, vol. 19, pp. 1261-1267, 2005.
[3] D. L. Mitchell and J. G. Speight, "The solubility of asphaltenes in hydrocarbon solvents," Fuel, vol. 52, pp. 149-152, 1973.
[4] M. Wang, Y. Hao, M. R. Islam, and C. C. Chen, "Aggregation Thermodynamics for Asphaltene Precipitation," AIChE Journal, Submitted.

Copyright © American Institute of Chemical Engineers. All rights reserved.

444086 Optimization of Layered Graphene Oxide Nanostructures through Controlling the Membrane Deposition Rate Towards Size-Sieving Separation

Abstract:

Single-layer graphene oxide (GO) flakes were coated on anodized aluminum oxide (AAO) substrate through a simple vacuum filtration process, the effect of the deposition rate on the nanostructures of obtained membrane was investigated though pure water flux and organic vapor diffusion, which indicated GO membrane assembled with slow deposition rate might have more ordered lamellar structure and, correspondingly, have smaller effective pore size. In contrast, under fast deposition rate, GO flakes assembled in a more random way. Therefore, through controlling the membrane deposition rate, the nanostructure of layered GO membrane could be optimized for size dependent separation. We found after thermal reduction, 100 nm GO membrane (prepared @ 0.02 nm/s) has effective pore size smaller than 6.1 Å, which exhibit great potential in n-hexane/2,2-dimethybutane (DMB) separation.

Copyright © American Institute of Chemical Engineers. All rights reserved.

444142 Robust Design of a Radiant Wall Burner for a High Temperature Environment Utilizing Computational Fluid Dynamics Simulations

Abstract:

Radiant Wall Burners are often utilized in Ethylene Cracking Furnaces. The conditions and fuel gases for a specific unit can vary considerably within the unit and across similar units. With burner air preheated to high temperatures, premixing the fuel can result in dangerous and destructive flashback, while complete nozzle-mixing of the fuel can result in burner instability at startup or turn-down conditions. These challenges can be solved through the generation of a robust radiant wall burner design, which is facilitated by executing Computational Fluid Dynamics simulations of many different geometric configurations.

Copyright © American Institute of Chemical Engineers. All rights reserved.

444161 Novel Microwave Synthesis Carbon Dots from PINE Bark and Its Application for Heavy Metal Sensing

Monday, April 11, 2016
Exhibit Hall E (George R. Brown)

Saheed Sanni, Chemistry, Vaal University of Technology, Vanderbijlpark, South Africa, Puthalapattu ReddyPrasad, Chemistry, Vaal University of Technology, Vanderbiljpark, South Africa and Augustine E. Ofomaja, Chemistry, Vaal university of Technology, Vanderbiljpark, WV, South Africa

NOVEL MICROWAVE SYNTHESIS CARBON DOTS FROM PINE BARK AND ITS APPLICATION FOR HEAVY METAL SENSING

Sanni S.O., ReddyPrasad P., Ofomaja A.E.

Biosorption and Waste water Treatment Research Laboratory, *Department of Chemistry, Faculty of Applied and Computer Sciences, Vaal University of Technology, P. Bag X021, Vanderbijlpark-1900, South Africa.*

INTRODUCTION

Microwave pyrolysis for the production of carbon has the advantages of being cheap, produces high yield of carbon and is absolutely a green synthetic method [1]. In recent times, the use of microwave pyrolysis have been extended to the preparation of carbon dots. Carbon nanodots (C-dots) falls under the class of carbon nanoparticles with size below 10 nm[2]. C-dots are carbon materials which processes high water solubility and strong fluorescence depending on their size, edge, shape, surface ligands, and defects. Since C-dots preparation can be achieved from difference carbonaceous sources and by varying a range of parameters during their synthesis, the properties of the final product may vary depending on these factors. C-dots have sp^2 characteristics and possess several functional groups on their surfaces that makes them highly hydrophilic and easily functionalized with various organic, polymeric, inorganic, or biological spices[3]. Several synthesis methods that have been explored includes electrochemical synthesis, combustion/thermal/hydrothermal/acidic oxidation, supported synthesis, microwave/ultrasonic, arc discharge, laser ablation/passivation, and plasma treatment[4]. C-dots processes excellent such as water dispersibility, chemical stability and photostability, ease of modification, excitation-dependent multicolor emission, low toxicity, and good cell permeability, great efforts have been paid to their potential applications in biosensing, bioimaging, drug delivery, and other biological related aspects.

METHODS

The carbon dots (C-dots) with green luminescence was synthesized from microwave assisted pyrolysis of pine bark waste our carbon source at 800, 1000 1200 W for 1 Hr. The dots produced were characterized using FTIR, TGA, XRD, UV-Vis and Raman spectroscopy. The dots were applied for heavy metal sensing.

RESULTS AND DISCUSSION

The functional groups present on the C-dots (**Fig.1**) includes -OH stretching band mode at 3413 cm^{-1}, C-H stretching at 2934 and 2845 cm^{-1}, stretching peaks of C=O, C=C at 1731, 1604 and 1437 cm^{-1}. The XRD pattern (**Fig 2a &b**) of produced C-dots depicts sharp peak at 20.9° and 26.8°, which is highly crystalline cores and possess amorphous nature of C-dots. This peculiar peak is ascribed to 002 of the graphitic carbon, which in line with inter-planar spacing of 3.77 Å with higher spacing than bulk graphite at 3.34 Å. The thermogravimetric and differential thermal analysis display (**Fig 3**) that the C-dots possess thermal stability more than 100 °C and mass losses occurs afterwards. The thermal analysis depicts 3 endothermic peaks at 331, 542 and 735 °C, which are ascribed to oxygenated groups such as carboxyl, hydroxyl, carbonyl and phenols. The optical properties of the synthesize C-dots were explored using UV-vis absorption and photoluminescence (PL) spectroscopy. The UV-vis spectra display (**Fig 4**)intense absorption peak in the range of 260 to 290 nm, which is ascribed to π-π^* transition of the aromatic sp2 domain and also presence of multiple polyaromatic chromophores. A shoulder peak at 330 nm which is ascribed to n-π^* transitions of C=O and structural defects of the C-dots. PL spectrum of the C-dots (**Fig 5a**)exhibits an intense maximum emission peak at 432 nm upon excitation at 330 nm, which also shows characteristic yellowish green color under sunlight and intense blue coloration UV lamp (365 nm). The PL properties of as prepared C-dots was varied at (**Fig5b**) different excitation wavelength from 300 to 405 nm. There was red-shifting in emission peak as the excitation peak was increased from 300 to 405 nm and also simultaneously decrease in fluorescent intensity, which is ascribed to quantum effect and surface defects. This behavior of the prepared C-dots is dependent on excitation-fluorescence, which is accordance with other report [13]. The quantum yield of synthesized C-dots was measured to be 17% at maximum excitation at 330 nm

Fig.6 shows the Raman spectrum of the PB-GQDs, which has been usually used to confirm the quality of the prepared PB-GQDs. There is two major Raman peaks at 1370 cm^{-1} (D band), 1580 cm-1 (G band). The peak of G band corresponds to the Sp^2 bonded carbon atoms in a two-dimensional (2D) hexagonal lattice and the peak of G band is associated with the vibrations of carbon atoms with dangling bonds in the termination plane of tangled graphite or glassy carbon. The D/G ratio band ratio is ID/IG: 0.68 is confirm the formation of graphene like structure from synthesized pine bark.

Fluorescence detection of Cu^{2+} ion

The detection of Cu^{2+} was performed at room temperature in phosphate buffer (PBS) solution pH 4.3. 1.0 mL of PB-CQDs solution was added into 0.5 mL of PBS buffer, followed by the addition of 1.0 mL different concentrations of Cu^{2+}. The fluorescence emission spectra were recorded after reaction for 5 min at room temperature (~28 °C).

Selectivity and sensitivity studies

Moreover sensitivity and selectivity is additional significant parameter to assess the performance of the sensing system. Consequently, we examined the fluorescence intensity changes in the presence of representative metal ions such as including Cu^{2+}, Hg^{2+}, Ca^{2+}, Al^{3+}, Ba^{2+}, Mn^{2+}, Ni^{2+}, Pb^{2+}, Co^{2+}, Zn^{2+}, Fe^{3+}, $Cs^{2+}$and Mg^{2+}, of each at a concentration of 15.0 μg mL^{-1} were reacted with a PB-CQDs solution. Fig. 7 A displays the relative change in PL intensity of PB-CQDs in the presence of various metal ions. Fe^{2+}, Mn^{2+} and Hg^{2+} ions cause the slight PL changes (defined as the relative change of PL intensity in 60–80% as compared to blank) which can be attributed to the nonspecific interactions between the carboxylic and/or amine groups and the metal ions. Among these metal ions, Cu^{2+} displays the strongest PL quenching effect on PB-CQDs, assigning to the special coordination between the Cu^{2+} ions and the phenolic hydroxyl and/or amine groups of PB-CQDs which has been widely used for the detection of Cu^{2+} ions or colored reactions in traditional organic chemistry. Such a specific fluorescence quenching effect may originate from the strong interactions between Cu^{2+} ions and the surface groups of PB-CQDs which transfer the photoelectrons from PB-CQDs to Cu^{2+} ions. These results clearly demonstrate that the PB-CQDs-based Cu^{2+} sensor is highly selective to Cu^{2+} over the other metal ions. Fig.8. Fluorescence responses of CCDs to the addition of different metal ions under acetate buffer (pH 4.3). The concentration of each metal ion is 15 μg $mL^{-1.}$ F_0 and F correspond to the fluorescence intensities of PB-CQDs at emission wavelength of 440 nm excited at 320 nm in the absence and presence of metal ions, respectively.

As shown in Fig. 3, the fluorescence intensity of PB-CQDs gradually decreased with the Cu^{2+} concentration. The fluorescence intensity displayed liner responses with the Cu^{2+} concentrations, with two linear ranges of 0.0–14.0 μg mL^{-1} and 15.0–25.0 μg mL^{-1} (Fig.3B). The linear equations are $F/F0 = 0.9827 - 0.8485 C_{Cu2+}$ (R = 0.993) and $F/F0$ = 0.7565–0.0033 C_{u2+} (R = 0.996), respectively.

REFERENCES

[1] W. B. Lu, X. Y. Qin, S. Liu, G. H. Chang, Y. W. Zhang, Y. L. Luo, A. M. Asiri, A. O. Al-Youbi, X. P. Sun, Economical, green synthesis of fluorescent carbon nanoparticles and their use as probes for sensitive and selective detection of mercury(II) ions, Anal. Chem. **2012**, 84, 5351-5357.

[2] J. Jiang, Y. He, S. Li, H. Cui, Amino acids as the source for producing carbon nanodots: microwave assisted one-step synthesis, intrinsic photoluminescence property and intense chemiluminescence enhancement. Chem Commun **2012**, 48, 9634–6.

[3] Y. Wang, A. Hu, Carbon quantum dots: synthesis, properties and applications, J. Mater. Chem. C **2014,** 2, 6921–6939.

[4] C. Liu, P. Zhang, X. Zhai, F. Tian, W. Li, J. Yang, et al., Nano-carrier for gene delivery and bioimaging based on carbon dots with PEI-passivation enhanced fluorescence, Biomater. **2012,** 33, 3604-13.

Figures:

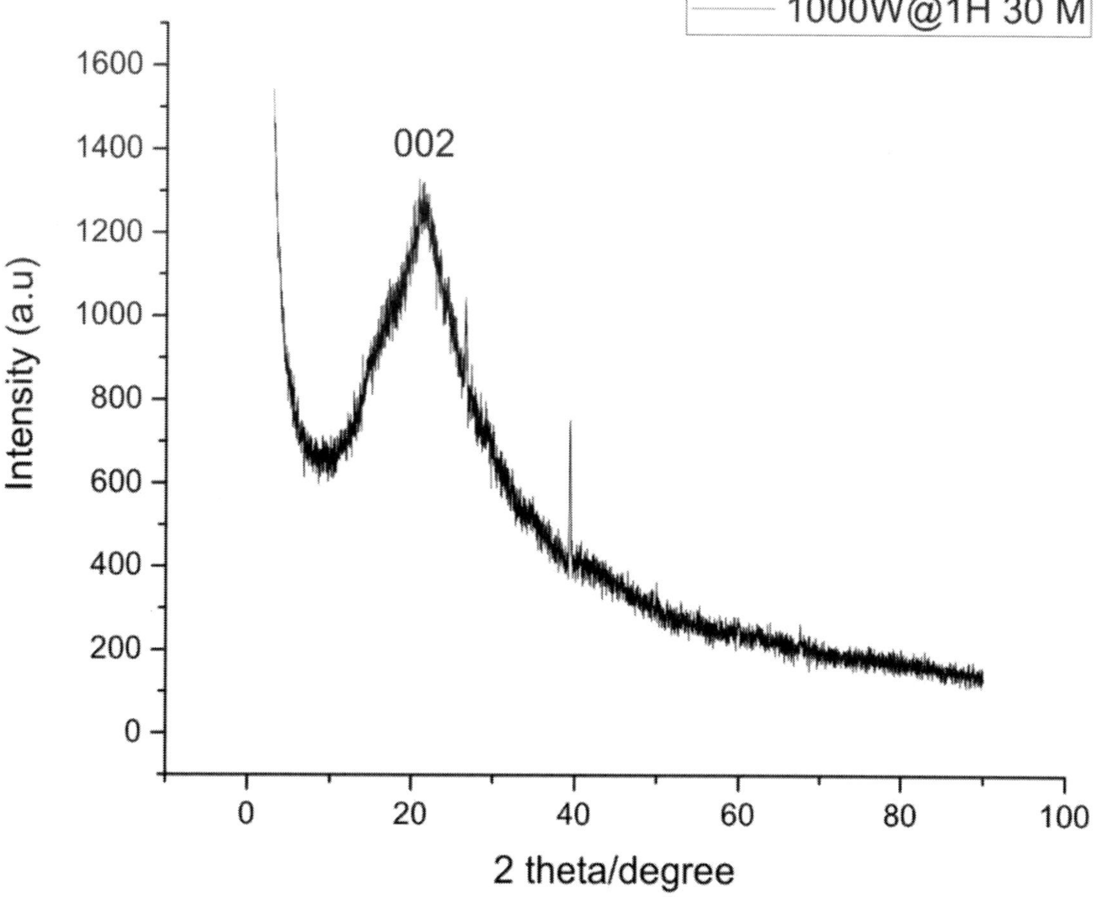
1000W@1H 30 M
1600
1400
1200
1000
800
600
400
200
0
Intensity (a.u)
002
0
20
40
60
80
100
2 theta/degree

Fig 1. FTIR spectra of as-prepared C-dots from Pine Bark

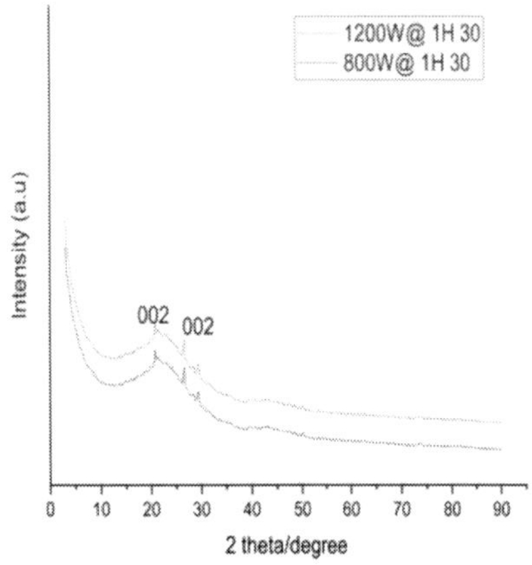

(a)

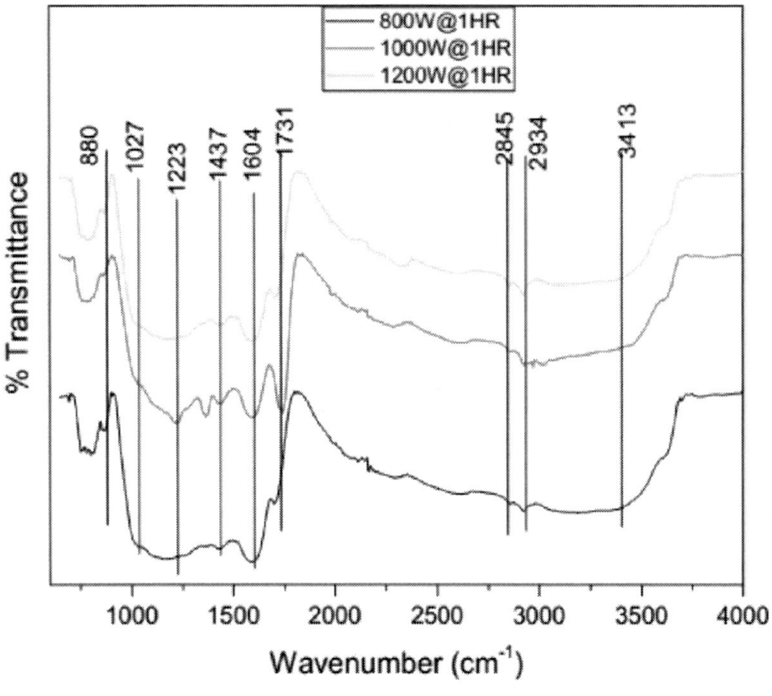

(b)

Fig. 2a & b. XRD pattern of C-dots.

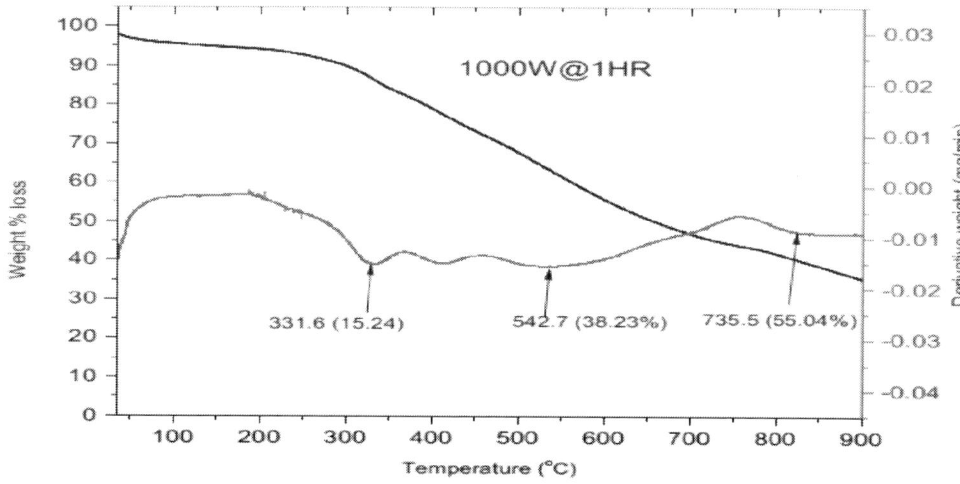

Fig 3. Thermal curves for C-dots under N_2 atmosphere.

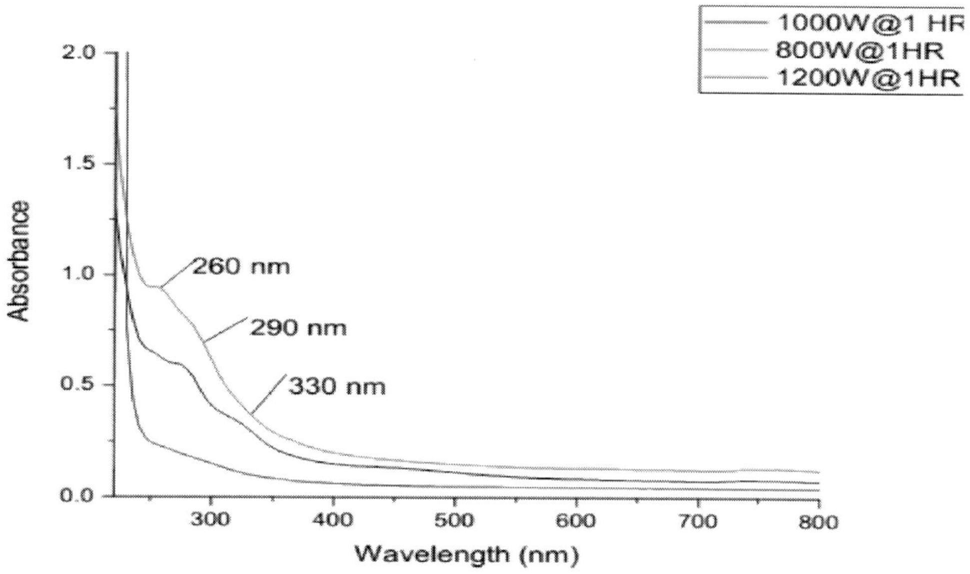

Fig 4. UV-vis absorption of C-dots.

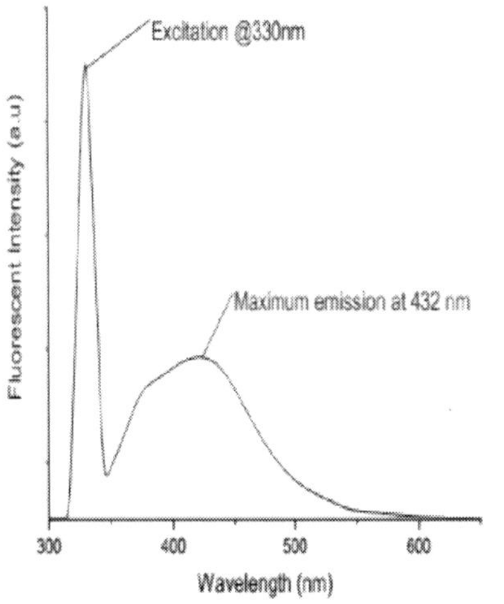

(a)

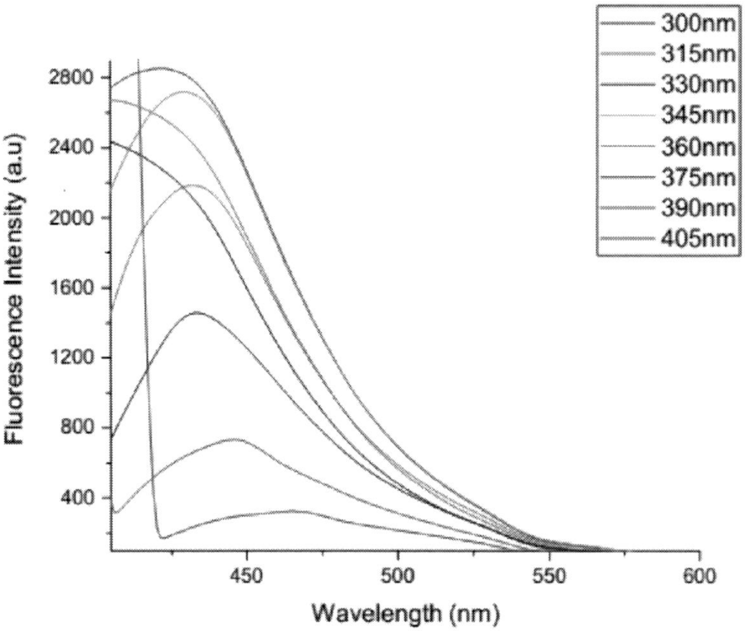

(b)

Fig 5a. PL excitation and emission wavelength (b) PL emission spectra at different excitation wavelength from 300 to 405 nm in 15 nm increments.

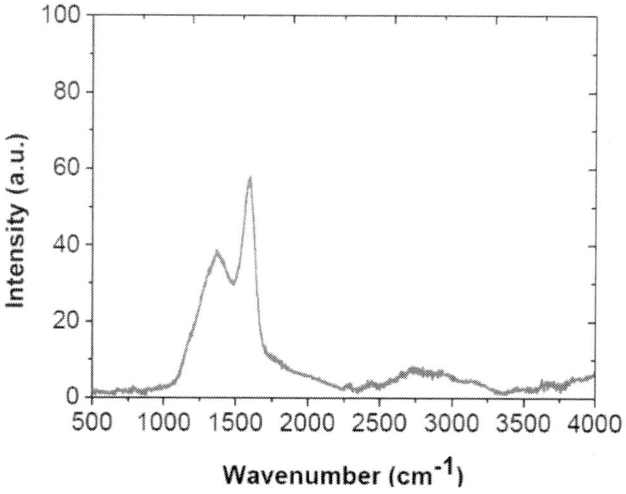

Fig. 6. Raman Spectra of PB-CQDs

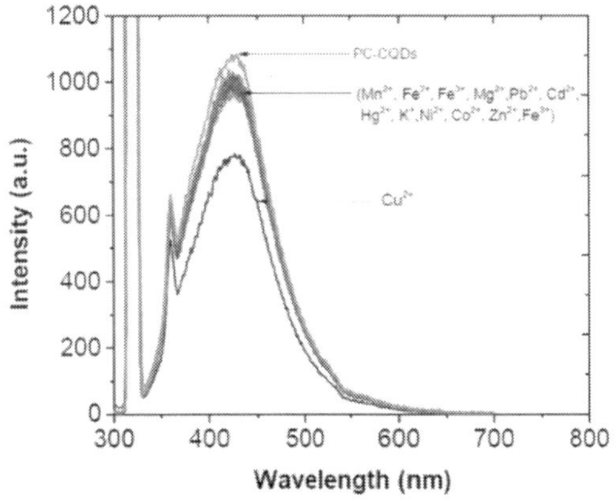

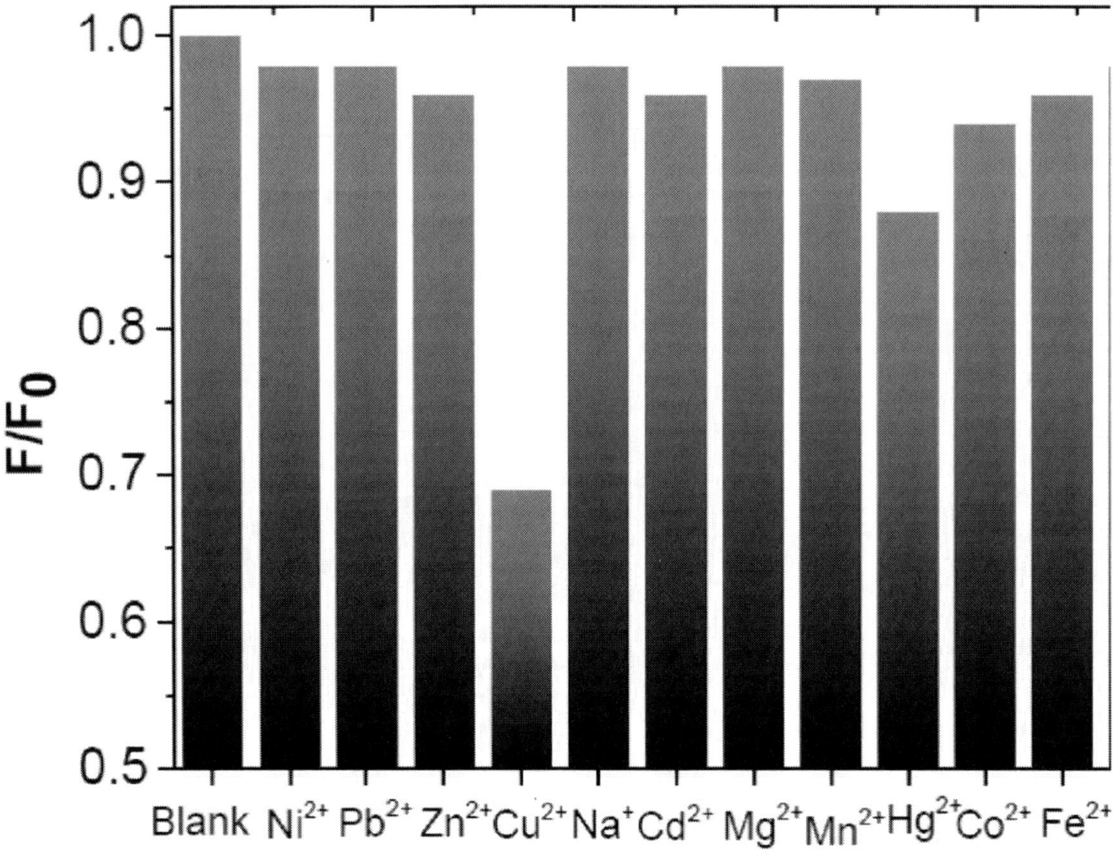

Fig.7. Fluorescence responses of CCDs to the addition of different metal ions under acetate buffer (pH 4.3).

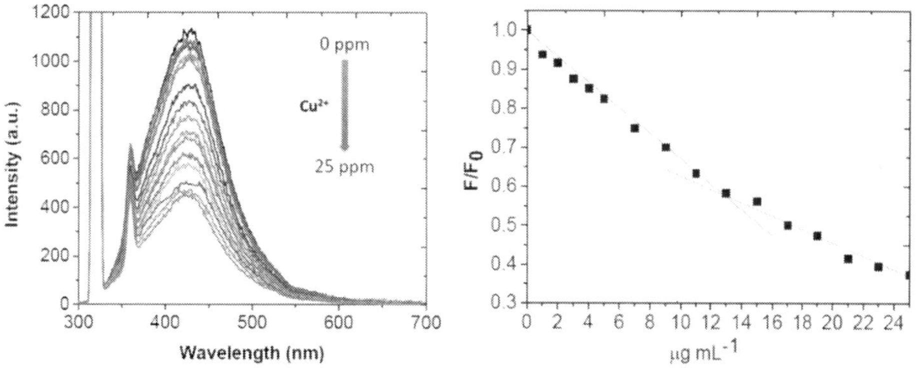

Figure 8. Fluorescence titrations of the y-CDs with Cu^{2+} from 0 to 25 μg mL^{-1}. (Influence of Cu^{2+} concentration on the fluorescence intensity of PB-CQDs (top to bottom: 0, 1, 2, 3, 4, 5, 7, 9, 11, 13, 15, 17, 19, 21, 23 and 25.0 μg mL^{-1})

444269 Functionalization of OH-Terminated Nanoparticles By Cyclic Azasilanes

Abstract:

Colloidal suspensions of functionalized nanoparticles have been suggested for enhanced oil recovery, formation damage remediation, flow assurance, and water treatment operations. Surface modified nanoparticles can also be used as tracers in reservoir studies, active media in thermal management fluids, and fillers in high performance elastomers. Conventional functionalization of nanoparticles with alkoxysilanes and chlorosilanes is limited by needs for catalytic amount of water for the reaction to proceed, requirement of multiple hours to fully saturate the surface, and release of by-products upon reaction completion. Monolayer deposition with a high density of functional groups represents a particularly difficult challenge.

Our ongoing studies indicate that application of cyclic azasilanes that have a strained ring structure with adjacent Si and N centers provides highly efficient functionalization of hydroxylated silica, alumina, magnetite, and diamond nanoparticles which is catalyst-free and by-product-free. The data on synthesis and characterization of functionalized nanoparticles as well as possible reaction mechanism of functionalization will be discussed.

Copyright © American Institute of Chemical Engineers. All rights reserved.

444093 New Development on Hoist Scheduling with the Relaxation of Cyclic Precondition

Abstract:

Multi-Stage Material Handling (MSMH) processes are designed to manufacture materials requiring a series of treatments and in industry, they are broadly used due to their capability to handle large work orders and compatibility with materials with different recipes. A typical example is electroplating process. A hoist handles the transportation of each job among units (stages) according to its processing recipe. The hoist movement plan is predetermined through optimization and this plan is repeated to perform continuity. Obviously, the ultimate goal is to maximize the productivity within each movement plan, which is equivalent to minimize the cycle time to finish one work order. The optimization process is generally called Cyclic Hoist Scheduling (CHS), since every movement plan is a cycle due to its repeatability. This precondition in term of hoist operation and unit usage is general for most cases. However, it could block the really optimal results.

In this study, this precondition has been relaxed for multi-capacity units, thus the unit usage can be more flexible. Thus, the real optimality can be found through CHS. To deal with the cycle connection issue from the relaxation of the precondition, a new module Cycle Connection Identification (CCI) is proposed. Obtained hoist schedule will be checked. Once a schedule with cycle connection problems is identified, a new cycle will be scheduled forcing the smooth connection with the first cycle. The objective of scheduling a new cycle is to match the beginning of the first cycle and the ending of the second cycle. If the two-cycle group is still with cycle connection issue, more cycles will be scheduled till the cycle group is free from cycle connection problems. Each cycle is called a sub-cycle and the constructed cycle group without cycle connection problems is called a full cycle. Note that in each sub-cycle, the work order and all transportation operations are completed.

In this paper, hoist scheduling models with/without such a precondition have been developed and proposed as mixed integer linear programming. The performance of the two models is tested through case studies and the difference is compared and demonstrated in detail.

Copyright © American Institute of Chemical Engineers. All rights reserved.

444359 Development of Novel Emulsified Acid Using Waste Oil

Abstract:

In carbonate reservoirs, hydrochloric acid (HCl) is commonly used for stimulation purposes. The main disadvantage for using HCl is its high and rapid reactivity allowing for less control on acid reaction with formation. In addition, HCl may react with the tubing causing severe corrosion and damages. Several slow-reacting acid systems, though HCl-based, have been developed to retard the HCl reaction and control its reactivity to allow deeper penetration in carbonate reservoirs before it breaks out and reacts with formation rock. However, the selection of the hydrocarbon phase of emulsified acid is critical to the creation and stability of the emulsified acid system. Diesel oil, crude oil and solvents such as xylene have been reported to be successful in the field.

In this work, alternative and less expensive source of oil is developed to replace the current practices. Emulsified acid is a water-in-oil emulsion where oil is the continuous phase and acid is the dispersed one. Industrial and filtered waste oil was used to prepare the emulsified acid of this work. Extensive work has been carried out to study the chemical composition of the waste oil, namely SARA and to select the proper emulsifier. Emulsions were prepared using 15% HCl with emulsifier concentration ranging from 0.5 to 2.0 vol% and 70:30 of acid-to-oil ratio. Droplet size distributions were measured and used to benchmark with other emulsions. Emphasis was given to the stability of the emulsion especially at high temperatures. In order to apprehend the effect of corrosion inhibitor on the emulsion stability, its concentration was varied (up to 0.3 vol%). Later, rheological analyses of the waste oil emulsion were compared to the commonly used diesel-based emulsified acid.

Results showed the potential use of waste oil to prepare high temperature emulsified acid (up to 100°C). The addition of corrosion inhibitor to the emulsified acid caused a negative effect on stability resulting in a less stable emulsion. On the other hand, addition of cationic emulsifier helped to formulate a more stable emulsion. Apart from that, the optimum concentration of the emulsifier was at 2.0 vol%. The rheological data confirms the stability of the new emulsified formulation at higher temperatures. This paper summarizes the findings of using waste oil emulsion and recommends it for field applications. The newly proposed emulsified acid introduces waste oil as a replacement of diesel.

The novelty of using waste oil will help to significantly reduce the expenses of formulating emulsified acid. Since waste oil is basically waste it has a good impact on the environment, that being said, instead of throwing it, the industry will be able to recycle it.

Copyright © American Institute of Chemical Engineers. All rights reserved.

444364 Molecular Simulation of Controlling Ions and Their Role in Low Salinity IOR for Carbonate Reservoirs

Abstract:

Carbonate geology represents around 50% of hydrocarbon-bearing reservoirs in the world and more than 70% in some areas such as the Middle East. Water flooding has not been very efficient in carbonate reservoirs due mainly to the low water wetness. Modifying the ionic composition and salinity of the injected seawater demonstrated that the oil recovery from these carbonate reservoirs could be improved. Although the underlying mechanisms are still a subject of debate, the improved oil recovery has been mostly attributed to the increased water-wetness of the rock through the interactions of specific ions like Ca^{2+}, Mg^{2+}, SO_4^{2-}, and Cl^- , which are present in seawater with the rock surface under favorable conditions

The study of hydrated gas-phase metal ions provides a link between intrinsic chemistry of the isolated ion and its chemistry in solution. The structures and binding enthalpies of a variety of gas-phase and solution mono- and divalent ion hydrates containing up to 12 water molecules have been studied computationally. A DFT and molecular dynamics simulations were employed to study hydrated structures of the form $[\mathbf{M}(H_2O)_n]^{2+}.mH_2O$ for calcium and magnesium and $[\mathbf{M}(H_2O)_n]^{+}.mH_2O$ for sodium, in which there are n water molecules in the first coordination shell and m water molecules in the second coordination shell. The hydration, dehydration, and interactions of metal ions with Cl^{-1} and SO^{-2}_4 have been studied to better understand the mechanism of low salinity IOR. Density functional (DFT) theory using the B3LYP functional, and 6-311++g** basis set has been used. The charge of the metal ion decreases monotonically as the number of water molecules increases in the complex. Extensive calculations of the energetic contribution of individual structural characteristics as well as the energetic trends in binding energy are developed, which will lead to better understanding of how low salinity water IOR works.

Keywords: Hydration, carbonate reservoir, DFT, binding energy, Low Salinity

Copyright © American Institute of Chemical Engineers. All rights reserved.

*** Formatting Issue - Best Available Paper/Graphic ***

444278 Insight into the Mechanism of Toxic Chromium (VI) from Wastewater Using Magnetite Coated PINE Powder

Abstract:

INSIGHT INTO THE MECHANISM OF TOXIC CHROMIUM (VI) FROM WASTEWATER USING MAGNETITE COATED PINE POWDER

Pholosi A., Ofomaja A., Naidoo E.B.

Biosorption and Wastewater Treatment Research Laboratory, Department of Chemistry,

Faculty of Applied and Computer Sciences, Vaal University of Technology, P. Bag X021, Vanderbiljpark, 1900 South Africa.

INTRODUCTION

A knowledge of the interaction between pollutant molecules or ions and active sites on an adsorbent surface is an important phenomenon in many industrial processes such as catalysis, adsorption and other separation techniques. The use of lignocellulosic materials for Cr(VI) adsorption has gained interest over the years (**Miretzky and Cirelli, 2010**). The adsorption properties of lignocellulosic materials has been attributed to acidic functional groups on their surfaces which can take part in ion-exchange, complexation, chelation and hydrogen bonding with pollutant species in solution (**Vithanage et al., 2015**). Two mechanisms for lignocellulosic adsorption of Cr(VI) have been proposed; (i) an electrostatic mechanism in which at low pH, the protonated acidic functional groups on the biomaterial carrying a positive charge attracts the negatively charged Cr(VI) species (**Demirbas, 2005**) and (ii) the adsorption-coupled reduction mechanism in which Cr(VI) is reduced to Cr(III) either in solution (Direct reduction) or after being adsorbed (Indirect reduction) by electron-donor groups of the biomaterial that have lower reduction potential values than that of Cr(VI) (**Gao et al., 2008**). The Cr(III) ions formed in solution may then form complexes with Cr(III)-binding groups on the biomaterial surface or released into bulk solution by repulsion between Cr(III) and other positively charged groups on the biomaterial (**Miretzky and Cirelli, 2010**).

METHODS

Sodium hydroxide treated pine coated with magnetite (NTP-NC) was prepared by co-precipitation method. The effect of solution pH was performed in a batch system in which 0.5 g of the adsorbent was contacted in six 250 cm^3 beakers containing 75 cm^3 of 75 mg/dm^3 Cr(VI) solutions set at pH 1, 2, 4, 6, 8 and 10 at 26 °C and agitated at 200 rpm for 2 h and the clear solution analysed for Cr(V/I) left using UV-Vis spectrophotometry at 460 nm

RESULTS

The results of the effect of initial solution pH on the adsorption of Cr(VI) onto NTP-NC is shown in Fig. 1. To confirm the operating mechanism, the change in $[H^+]$ concentration at each initial pH, the oxidation/reduction potential before and after adsorption and the amounts of Total Cr, Cr(VI)and Cr(III) present in solution at each pH was monitored for NTP-NC and the results displayed in Fig1 . The change in $[H^+]$ concentration ($\Delta[H^+] = [H^+]_{final} - [H^+]_{initial}$) at the end of the adsorption process for each initial solution pH was recorded and plotted against the initial solution pH. The results showed that the change in $[H^+]$ concentrations at the end of the adsorption process where higher for lower initial solution pH's (where lower amounts of Cr was found in solution) than for higher initial solution pH's (where higher amounts of Cr was found in solution) and the change in $[H^+]$ concentration became small and

*** Formatting Issue - Best Available Paper/Graphic ***

almost constant after initial solution pH 4. This is consistent with the assumption that the removal of Cr from solution is associated with $[H^+]$ ion consumption and that as the concentration of $[H^+]$ ions reduces in solution, the Cr removal reduces. Cr(VI) ion in acidic solution has a positive redox potential value (above 1.3 V at standard conditions) and in the presence of an electron donor becomes unstable and can be reduced to Cr(III) **(Park et al., 2008)**. The results show that the initial ORP values (before adsorption) for the Cr solutions at pH's of 1, 2, 4, 6, 8, and 10 were 339, 280, 165, 55, -53 and -172 mV for NTP-NC respectively. The positive values of ORP at lower solution pH's indicates higher possibility for reducing Cr(VI), while the lower and negative ORP values indicates lower possibility for reducing Cr(VI). On addition of the adsorbents into the solutions, a change in ORP were observed . The values of ORP at the end of the adsorption process became 330, 268, 30, 9 -29 and -61 for NTP-NC respectively. A reduction in ORP at the end of the adsorption stage signifies that reduction has occurred in the system. **Yu et al. (2014)** showed the amount of Cr(VI) removed from aqueous solution in contact with zero-valent iron (nZVI) could be related to the change in ORP and pH of the solution. The authors observed a decrease in ORP from 400 to 100 mV and a corresponding 10 unit pH increase as Cr(VI) was removed from solution.

Finally, analysis of the prepared Cr solutions at the various solution pH's showed that the concentration of Total Cr obtained from ICP and that of Cr(VI) using UV-vis measurements were very similar, suggesting that Total Cr was almost equal to Cr(VI) in solution. After contacting the various solutions with the adsorbents, the concentration of Total Cr in the pH 1 solution was observed to have reduced to 21.56 mg/dm^3 for NTP-NC, while the Cr(VI) concentrations was 10.66 mg/dm^3 for NTP-NC. With solution pH 2, the concentration Cr(III) left in solution was lower than that of solution pH 1 for both samples. The reason is that at pH 1 there is an abundance of H^+ ions in solution which accumulates on the functional groups on the adsorbent making the surfaces of the adsorbent positively charged and repelling away the positively charged Cr(III) ions from the adsorbent surfaces **(Suksabye et al., 2009)**. At solution pH 2, it is believed that there is less accumulation of H^+ ions at the adsorbent surfaces which led to reduced electrostatic repulsion and increased adsorption of Cr(III) from solution. This increase in Cr(III) adsorption is also confirmed by the reduction in Total Cr in solution between solution pH 1 and 2. The concentrations of Cr(VI) left in solution at pH 2 was higher than at pH 1 for NTP while for NTP-NC the Cr(VI) concentrations were lower at pH 2. For solution pH's higher than 2 and up to 10, the residual Cr(III) in solution becomes very low and almost constant. This is consistent with the small change in $[H^+]$ in solution which is almost constant at this pH range.

FTIR Evidence for the adsorption of Cr(III)

When Cr(III) is formed in solution via adsorption-coupled reduction mechanism it may undergo hydrolysis and/or complexation depending on the solution pH and the total Cr(III) concentration. Hydrolysis may lead to the formation of various Cr(III) species with an increase in solution pH which gives rise to a reduction in its solubility and increasing the chances of precipitation of the Cr(III) species. At solution pH 1 – 4, Cr^{3+} and $CrOH^{2+}$ are the major species which can be adsorbed by negatively charged ligands on the biomaterial surface. On the other hand, at solution pH range of 4 – 10, Cr(III) species (hydroxo complex), such as $CrOH^{2+}$, and $Cr(OH)_3$, are the dominant forms in the aqueous system and at high pH is readily transformed into soluble complex, . When NTP-NC was contacted with Cr(VI) solution for 2 hr, the final solution pH was increased to pH 2.53. The FTIR spectra for NTP-NC (Fig. 2) shows the presence of characteristic Fe-O peaks of bulk Fe_3O_4 at 553.26 cm^{-1}, 564.36 cm^{-1}, 570.21 cm^{-1} and 576 21 cm^{-1}. Functional groups associated with the pine includes the broad band at 3250.99 cm^{-1} representing –OH of cellulose, 2905.17 cm^{-1} representing C-H stretching, while the peaks at 1627.69 and 1602.87 cm^{-1} shows the COO^- and C=O bonds respectively. After 2 hr contact with Cr(VI) solution, the peak at 564.34 cm^{-1} was observed to have split into several new peaks (Fig. 3b (ii)). These emerging peaks are characteristic of Fe-O bonds of maghemite (γ-Fe_3O_4), for example the characteristics peaks of maghemite was observed at 595.21, 685.65 cm^{-1} and 582.70, 634.91 cm^{-1} suggesting a change in the iron oxide phase (Fe_3O_4 to γ-Fe_2O_3) due to the conversion of Cr(VI) to Cr(III) by magnetite nanoparticles. According to **Peterson et al. (1997)**, Cr(III)(hydr)oxide produced is precipitated on the magnetite surface. In this study, α-Cr_2O_3 peak were observed at 604.63 cm^{-1} indicating that Cr(III) was formed during the adsorption process. Finally it was also observed that the carboxylate group at 1635.55 cm^{-1} was shifted to 1637.73 cm^{-1} while a new peak was observed at 1737.72 cm^{-1} suggesting that complexation between Cr(III) and Carboxylate ions also occurred.

REFERENCE

Demirbas, A. Energy Source 27 (2005) 1449–1455.

Miretzky, P., Cerelli, A.F. A review. J. Hard. Mater. 180 (2010) 1-9.

Vithanage, M., Rajapaksha, A.U., Ahmad, M., Uchimiya, M., Dou, X., Alessi, D.S., Ok, Y.S. J. Environ. Manage. 151 (2015) 443-449.

D. Park, C.K. Ahn, Y.M. Kim, Y.-S. Yun, J.M. Park, J. Hazard. Mater. 160 (2008) 422–427.

P. Suksabye, A. Nakajima, P. Thiravetyan, Y. Baba, W. Nakbanpote.. J. Hazar. Mater. 161 (2009) 1103–1108.

R.F., Yu, F.H., Chi, W.P., Cheng, J.C. Chem. Eng. J. 255 (2014) 568–576

M.L. Peterson, A.F. White, G.E. Brown, G.A. Parks Environ. Sci. Technol. 31 (1997) 1573–1576.

Fig. 1 Effect of solution pH on the adsorption of Cr(VI)

*** Formatting Issue - Best Available Paper/Graphic ***

Fig, 2: FTIR of NTP-NC

Copyright © American Institute of Chemical Engineers. All rights reserved.

444280 Copper-Phyllosilicate Core-Shell Nanocatalysts with Balanced Active Sites for Carbon-Oxygen Hydrogenolysis Reactions

Abstract:

Copper-Phyllosilicate Core-shell Nanocatalysts with Balanced Active Sites for Carbon-Oxygen Hydrogenolysis Reactions

Hairong Yue, Weimao Fang, Changmin Liu, Jiankang Ying

College of Chemical Engineering, Sichuan University, Chengdu, China

Email: hryue@scu.edu.cn

Abstract

Hydrogenolysis of carbon-oxygen (C¨CO) bonds (e.g., esters, ethers, furfural, and CO_2) has emerged as a versatile synthetic tool in organic methodology, as it could produce a variety of products (e.g., chemicals, fuels, and polymers). Copper-based catalysts have been intensively explored for hydrogenation reactions as the copper sites account for the selective hydrogenation of carbon-oxygen bonds and relatively inactive for the hydrogenolysis of carbon-carbon bonds. Work on understanding the active sites of copper catalysts for hydrogenation reactions indicted that both Cu^0 and Cu^+ species were crucial to the activity of Cu-based catalysts. However, researchers were not able to establish a relationship between activity and Cu^0/Cu^+ active species since the conventional Cu-based catalysts is the deactivation by metal particle growth and unstable surface Cu^0 and Cu^+ active species in the strong reductive H_2 and oxidizing carbon-oxygen atmosphere.

We renctly present on efficient approaches for fabrication of a series of copper-phyllosilicate core-shell nanocatalysts and nanoreactors with balanced and stable Cu^0 and Cu^+ active species. We chose the hydrogenation of dimethyl oxalate, ethylene cyclicarbonate and as probe reactions for a better understanding of properties of active sites and the structure-activity relationship. The results indicated a synergy between copper and the oxide components and the balanced surface Cu^0 and Cu^+ species can greatly improve the catalytic C-O hydrogenolysis performance of the catalysts. These nanocatalysts with balanced and stable Cu^0 and Cu^+ active species, confinement effect, intrinsic high surface of Cu^0 and Cu^+ and unique tunable tubular morphology, has potential applications in the high-temperature hydrogenation reactions.

Keywords: core-shell nanocatalysts, Hydrogenation reaction; copper catalyst

References

1. Yue, H.; Zhao, Y.; Ma, X.; Gong, J.: Ethylene glycol: properties, synthesis, and applications. *Chem Soc Rev* 2012, 41, 4218-4244.

2. Gong, J.; Yue, H.; Zhao, Y.; Zhao, S.; Zhao, L.; Lv, J.; Wang, S.; Ma, X.: Synthesis of Ethanol via Syngas on Cu/SiO2 Catalysts with Balanced Cu^0-Cu^+ Sites. *J Am Chem Soc* 2012, 134, 13922-13925.

3. Zhao, S.; Yue, H.; Zhao, Y.; Wang, B.; Geng, Y.; Lv, J.; Wang, S.; Gong, J.; Ma, X.: Chemoselective synthesis of ethanol via hydrogenation of dimethyl oxalate on Cu/SiO_2: Enhanced stability with boron dopant. *J Catal* 2013, 297, 142-150.

4. Yue, H.; Ma, X.; Gong, J.: An Alternative Synthetic Approach for Efficient Catalytic Conversion of Syngas to Ethanol. *Acc Chem Res* 2014, 47(5):1483-1492

5. Yue, H.; Zhao, Y.; Zhao, L.; Lv, J.; Wang, S.; Gong, J.; Ma, X.: Hydrogenation of dimethyl oxalate to ethylene glycol on a Cu/SiO2/cordierite monolithic catalyst: Enhanced internal mass transfer and stability. *AIChE J* 2012, 58, 2798-2809.

6. Yue, H.; Zhao, Y.; Zhao, S.; Wang, B.; Ma, X.; Gong, J.: A copper-phyllosilicate core-sheath nanoreactor for carbon¨Coxygen hydrogenolysis reactions. *Nat Commun* 2013, 4, 2339.

Copyright © American Institute of Chemical Engineers. All rights reserved.

Stimuli-responsive Rubber in Oil and Gas Applications

Rostyslav Dolog, Sankaran Murugesan, and Valery Khabasheskii

Baker Hughes, 14990 Yorktown Plaza Dr, Houston, Texas, 77040, USA

ABSTRACT

A problem that exists today in the oil & gas industry is the loss of mechanical properties by reactive rubber seals due to swelling, and the limited space in downhole well applications to accommodate larger swellable packing elements.

This paper presents cost-effective, reinforcing, stimuli-responsive fillers, such as certain classes of fly ash, for use in rubber composite materials in various oil and gas applications. Apart from cost effectiveness, these fillers can also enable stimuli-responsive effects, as well as improved mechanical properties (e.g. modulus) for rubber composites when swelled in water. Fly ash (Class C) is a common component of Portland cement which was used in the literature to achieve a stimuli-responsive effect when added to Hydrogenated Nitrile Butadiene Rubber (HNBR) as filler. Exposure to water brines can cause a filler to promote an increase in pH which enables the hydrolysis of HNBR nitrile groups to carboxyl groups, thus creating an ionomer. This process is followed by neutralization of the new ionomer with metal ions, and formation of ionic crosslinks, further improving mechanical properties of the rubber.

This paper demonstrates that incorporation of fly ash as a reinforcing filler allows development of swellable seals that can retain, or even increase, their modulus and stiffness upon swelling. These characteristics allow avoiding the loss of mechanical properties by seals due to swelling and also eliminating the need for large swellable packing elements that occupy additional space downhole.

KEYWORDS: Elastomers, stimuli-responsive, fly ash, composites

INTRODUCTION

There is a growing demand for high performance elastomers in the oil & gas industry. Rubber compounds are used for various sealing applications, packer elements, stators for pumps, etc. In general, rubber formulations are comprised of elastomer matrix, curing system, antioxidants, plasticizers, along with reinforcing filler, which provides the compounds with their strength. The reinforcing filler is usually represented by carbon black and, to a lesser extent, silica. Along with other factors, filler mechanical properties and concentration generally define hardness, modulus and tensile strength of the elastomer compound. Throughout the years, many different materials have been tried as alternative fillers for rubber composites, including nanomaterials (CNTs, graphene, nanofibers), recycled rubber particles, and ash/fly ash. In the case of ordinary carbon black, the filler serves not only for reinforcement, but also as a cost reducer, because raw elastomer is often a more expensive component. Mechanical properties of the composites are affected by the aspect ratio of the filler, quality of the dispersion, and interface between blend components. The latter determines the effectiveness of load transfer from polymer matrix to the reinforcing filler. High performance compounds can be designed by incorporating well dispersed fine particles with a high aspect ratio and good compatibility with the polymer matrix.

In this context, the fly ash is a cheaper, low reinforcing filler alternative to carbon black, which will help reduce cost of the compounds for "non-demanding" sealing applications requiring large amounts of raw materials. Fly ash is produced in large quantities as a byproduct of coal burned at power plants, and can be considered as a recycled material. Fly ash costs several cents per pound, while carbon black, depending on the grade, is usually one order of magnitude more expensive. Hemmings et al[1] demonstrated that fly ash can be used as filler for polymeric composites with enhanced mechanical properties. Adoption of this technology in the oil and gas industry could change the design approach for reactive rubber compounds. Further functionalization of fly ash through a cost-efficient process could ensure compatibility with the rubber matrix, while promoting filler dispersion and ensuring an efficient load transfer. Such functionalization's include, but are not limited to, surface modification with fatty acids, dicarboxylic acids or phosphoric acid. The reasoning for surface modification is to attach functional groups with aliphatic chains to fly ash particles to make them compatible with aliphatic rubber chains, or to attach acidic groups on their surface that can interact with nitrile groups of HNBR. Initial attempts towards chemical functionalization of fly ash have been made by Seal et al for treatment of offshore oil spills.[2]

Apart from cost effectiveness, composites with fly ash can show stimuli-responsive properties. It has been shown that adding Portland cement to the HNBR matrix improves its mechanical properties in the swollen state by water uptake.[3] Exposure to water causes Portland cement to grow in volume and mass, and promotes an increase in pH which enables the hydrolysis of HNBR nitrile groups to carboxyl groups, thus creating an ionomer. This process is followed by neutralization of the new ionomer with metal ions and formation of ionic crosslinks, further improving mechanical properties of the rubber. Fly ash (Class C) is a common component of Portland cement, which contains a large variety of metal oxides, and is capable of increasing the

---

[1] Filler Comprising Fly Ash For Use In Polymer Composites. Hemmings et.al., US 6,916,863 B2, July 2005.

[2] Hydrophobic High Surface Area Zeolites Derived from Fly Ash for Oil Spill Remediation, Seal et.al., Environ. Sci. Technol., 2013, 47 (11), pp 5843–5850

[3] Stimuli-Responsive Cement-Reinforced Rubber, Musso et. al., ACS Appl Mater Interfaces. 2014 May 14;6(9):6962-8.

pH of the water solution by hydrolysis, which has a potential to achieve a stimuli-responsive effect when added to HNBR as filler.

Degradation of tensile properties is a common problem for swellable packers and other sealing elements. By selectively using certain classes of fly ash as effective reinforcing filler, one can develop swellable seals that can retain or even increase their modulus and stiffness upon swelling. Therefore, there is a potential to minimize or even avoid the loss of mechanical properties by water swellable reactive packing elements due to swelling. This provides an opportunity to decrease the size of these elements, which are often represented by rather long sections, and hence occupy a significant amount of space downhole. In this paper we have studied the reinforcing effects of fly ash fillers with HNBR compounds as well as stimuli responsive properties of these compounds.

EXPERIMENTAL SECTION

Rubber compounds T1, T2, T3, and T4 are elastomers based on a custom-built peroxide curing system. Formulation F2 was prepared by adding 15 wt. % of fly ash class C to compound F1, which is a commercial swellable NBR sulfur-cured compound.

T series compounds were prepared using a C.W. Brabender® internal mixer operated at 60 RPM for 20 minutes at ~160°F and then molded into sheets and cured at 340°F for 30 minutes in a hot press. Compound F2 was prepared by mixing fly ash with compound F1 in the Brabender mixer for 6 minutes at 60 RPM at ~100°F. Compound F1 is a commercial water swellable formulation that was used as received. F series compounds were cured at 300°F for 60 minutes in a hot press.

Tensile properties of rubber samples were tested using dog bone specimens, according to the ISO37 standard, at room temperature and at 350°F using an MTS® testing system.

Swelling tests in 3% KCl brine and LVT 200 oil were performed to assess the effect of fillers on the chemical resistance of elastomers. Standard one-inch rubber buttons were tested at 200°F at ambient pressure for 6 days. Thermal properties were studied using TGA Q50 and DSC Q2000 instruments.

To analyze the stimuli-responsive effect introduced by fly ash fillers, rubber buttons were aged for 10 days in 3% KCl in the oven at 200°F and then removed from the solution, dried for 30 minutes in the oven and then for 3 hours at room temperature. Samples were later tested in the MTS testing system in compression mode.

Viscoelastic properties of stimuli-responsive samples, based on water swellable compounds, were accessed using the RSA G2 dynamic mechanical analysis (DMA) instrument. This experiment was conducted in a temperature ramp mode of 2°C/minute at 1Hz frequency and 0.2% amplitude strain. A Nicolet 6700 Fourier Transform Infrared (FTIR) spectrometer was used to confirm neutralization of carboxylic groups by metal cations.

RESULTS AND DISCUSSION

The reinforcing properties of fly ash were evaluated at room temperature (Table 1) and at 350°F (Table 2) and compared with N550 carbon black. Sample T3, reinforced with fly ash, demonstrated inferior tensile properties and hardness compared to a compound filled with carbon black. However, at room temperature, fly ash was capable of delivering a reinforcing effect to the rubber compound without sacrificing its elasticity, e.g. the ultimate elongation of elastomer compound remained virtually unchanged after incorporation of 50 parts per hundred rubber (PHR) of fly ash, while its modulus and tensile strength have increased. Therefore, potentially, one can substitute carbon black with high loadings of fly ash in certain applications, thus further decreasing the cost of the rubber product. The summary of mechanical properties is provided in Table 1, E_x denotes modulus of a compound at certain strain (x), i.e. stress at strain.

Sample	N550 CB (PHR)	Fly ash Class F (PHR)	Tensile strength (psi)	E_{25} (psi)	E_{50} (psi)	E_{100} (psi)	Elongation (%)	Hardness Shore A
T1	0	0	829	71	97	121	472	48
T2	50	0	3402	165	273	744	290	68
T3	0	50	1176	94	127	166	467	54

Table 1. Room temperature tensile properties and hardness of elastomer samples

The relatively low increase in modulus and hardness of T3 compared to T1 at room temperature can be explained by a low aspect ratio of fly ash and weak interactions between blend components. The latter problem can be solved by functionalization of fly ash particles.

The general data trends from the high temperature tensile tests closely resemble the results of the tensile test conducted at room temperature. However, at high temperature, rubber specimens break at much lower elongation, which leads to a great decrease in the measured tensile strength. This behavior is typical for elastomers and limits their application to a certain maximum service temperature and is also due to the fact that high temperature tests are conducted at conditions that push the limits of the material. The modulus values at 15% (E_{15}) and 20% (E_{20}) strain are not provided herein, because half of the tested specimens were broken before they reached such strains.

The superiority of carbon black as an efficient filler became even more pronounced at high temperature. However, incorporation of fly ash doubled the tensile strength and increased the elongation at break by one-third when compared to the unfilled elastomer.

Sample	E_5 (psi)	E_{10} (psi)	E_{15} (psi)	E_{20} (psi)	Tensile strength (psi)	Elongation (%)
T1	31	46			91	23
T2	61	130	194	261	832	39
T3	42	73	98	119	173	30

Table 2. Summary of high temperature (350°F) tensile data for elastomer samples

Thermal properties of samples T1-T3 were assessed with Differential Scanning Calorimetry (DSC) and Thermogravimetric analysis (TGA). TGA allows evaluating thermal stability of samples and to verify the actual composition of a sample from the ash content. The results of the DSC and TGA analysis are summarized in Table 3. During TGA, samples were heated to 600°C at 10°C/minute rate in a nitrogen (N_2) environment, and subsequently heated at 10°C/min to 600 °C in air to burn off carbon-based fillers. This allows decomposing of organic matter first by applying heat, and then burn the remaining carbon-based material in air atmosphere.

The initial degradation of the polymer started at 200°C and progressed to complete decomposition of aliphatic chains in the temperature range of 420-500°C with a highest rate of decomposition at ~460°C. Sample T2 incurred the second rapid mass loss after 600°C due to burning of the carbon black, which comprised most of the ash content at 600°C in a N_2 environment. The DSC experiment demonstrated the glass transition temperature of the elastomer remained largely independent of the type of filler - the respective DSC heating scan was conducted with a 10°C/minute temperature ramp after precooling the sample to -50°C.

Sample	W_{600} (%)	W_{800} (%)	$T_{deg}1$, °C	$T_{deg}2$, °C	T_g, °C
T1	2.4	2.4	452.8	-	-29
T2	29.8	2.7	458.7	616.5	-29
T3	32.8	32.7	447.3	-	-28

Table 3. TGA/ DSC results for elastomer samples

Table 3 contains ash contents (W) at 600°C and 800°C, degradation temperatures (T_{deg}) which represent the temperature of the highest rate of decomposition during TGA experiment, and glass transition temperature (T_g) of the samples, as determined by DSC scan.

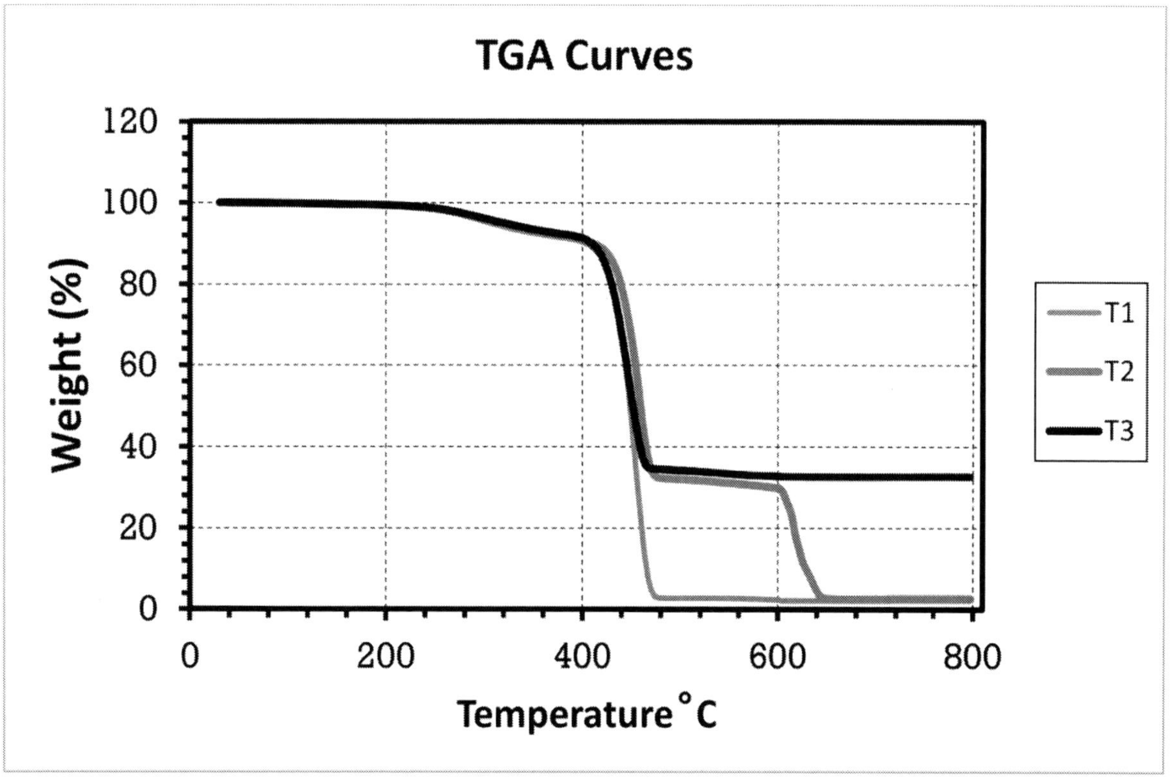

Figure 1. TGA Curves for samples T1, T2, and T3

Sample \ Testing fluid	3% KCl brine	LVT 200	Specific gravity
T1	0.6	13.0	0.96
T2	0.3	11.6	1.13
T3	0.8	9.1	1.18

Table 4. Volume increase (%) of elastomer samples due to swelling for 6 days at 200°F

Musso et. al. developed a stimuli-responsive HNBR composite reinforced with Portland cement that was increasing its stiffness and modulus when swelled in water[3]. Fly ash (Class C) is a common component of Portland cement, which contains the components necessary to achieve a similar effect. Therefore, one can avoid the loss of mechanical properties by water swellable packing element seals upon activation. To test this hypothesis, we studied the effect of aging in water brines on the properties of elastomers with fly ash fillers.

Rubber samples were molded and cured into standard size rubber buttons (cylinders of ~0.25" height and ~1" diameter). These buttons were aged for 10 days in 3% KCl brine in the oven at 200°F, removed from the solution, dried for 30 minutes in the oven and for 3 hours at room temperature. The swelling of both compounds was close to ~1% by volume (Table 4). Samples were later tested in an MTS machine in compression mode. Each button was compressed to 70% strain, buttons remained intact during the experiment, and visually all appeared to have a certain level of compression set at the end of the experiment. The results of the experiments are summarized in Table 5 and Figure 2. Only one button was tested for each sample. The provided compression modulus values (C_n) represent values of stress at a percentage strain (n), e.g. C_5 is stress at 5% compression strain.

Compound T4 was reinforced by fly ash Class C, which contains larger weight fraction of CaO than fly ash Class F (~25% vs ~10-15% respectively) and is expected to produce a higher pH and closely resemble cement that was used to prepare stimuli-responsive rubber in the original paper by Musso et.al.[3] The results demonstrate that exposure to water (3% KCl) at high temperatures can improve the modulus for the compounds filled with fly ash Class C (T4), while the control sample (T1) showed the opposite behavior. After aging, the compressive modulus at 50% strain (e.g. stress at 50% strain) of control sample T1 decreased by 19%, as compared to the unaged sample, while compressive strength of sample T4, reinforced by fly ash, increased by 15%. The difference in modulus values is even more drastic at 70% compression strain (-35% and 31%, respectively). This is in agreement with the hypothesis stated earlier.

Sample	Filler	Compressive modulus (psi)				
		C_5	C_{10}	C_{25}	C_{50}	C_{70}
T1	None	19	42	127	468	1810
T1 swollen	None	16	36	109	380	1185
T4	Fly Ash class C (50 PHR)	21	46	134	434	1628
T4 swollen	Fly Ash class C (50 PHR)	29	57	153	501	2125

Table 5. Summary of compressive modulus properties after 3% KCl soaking and 200°F drying

To further investigate the stimuli-responsive property of fly ash filler, the latter was incorporated into a water swellable compound, which provided for significantly more interaction of filler with water to occur. Sample F2 comprises unmodified commercial water swellable NBR formulation F1 mixed with 15 wt.% of fly ash Cass C. Sample F2 was processed into a 0.04" slab, then a rectangular sample was cut from the slab and swollen in 3% KCl solution until 100% volume increase. The F2 sample was then dried overnight in the oven, which lead to de-swelling of the sample due to solvent evaporation. Mechanical properties of the F2 sample were assessed with DMA prior to the swelling test, in a swollen state, and after drying - the data summary is shown on Figure 3.

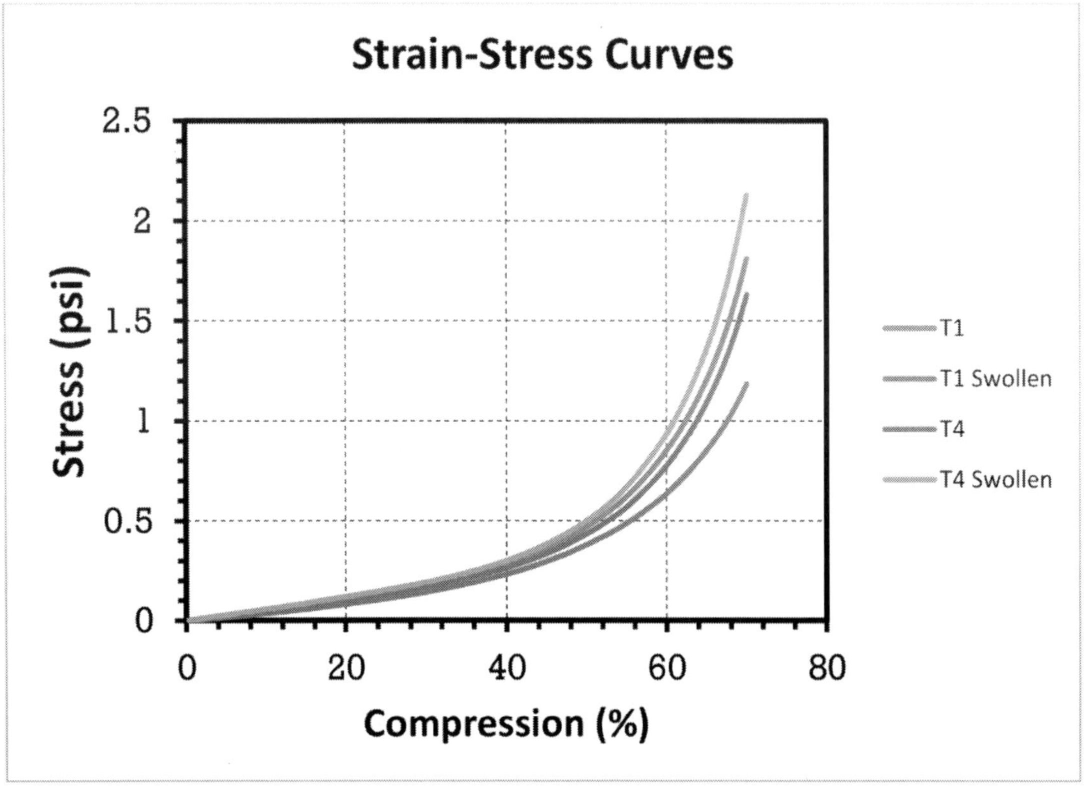

Figure 2. Stress-strain curves

The storage modulus of sample F2 after swelling and subsequent drying was about an order of magnitude higher than that of original sample F2 measured prior to swelling experiment. The storage modulus of the dried F2 sample was also less sensitive to the increasing temperature, contrary to the weakening original sample F2. The increase in the storage modulus of swollen F2 with temperature is simply due to the drying of the sample. The obtained result demonstrated that fly ash can essentially act as a stimuli-responsive material.

The significant improvement in the modulus is believed to be due to the neutralization of the swellable component of the commercial F1, and potential hydrolysis with subsequent neutralization of NBR elastomer. This is confirmed by the shift in T_g to a lower temperature range after drying as illustrated in Figure 3.

Carboxylic groups were detected in the original sample F2 prior to swelling, hence, they cannot be attributed solely to hydrolyzed HNBR, but rather to the super-absorbent polymer present in water swellable rubber compound (see Figure 4). FTIR confirms neutralization of carboxylic groups because a significant increase of the peak near ~1540 cm^{-1} and a lowering of peak intensity at ~1720 cm^{-1} have been observed. This neutralization is believed to promote the formation of ionic clusters in the compound with fly ash considered to be the source of cations. The neutralization of carboxylic groups could as well be attributed to K^+ cations present in the KCl brine solution, however, in that case, the respective peak in FTIR spectra would appear at higher wave numbers, approaching ~1600 cm^{-1}.

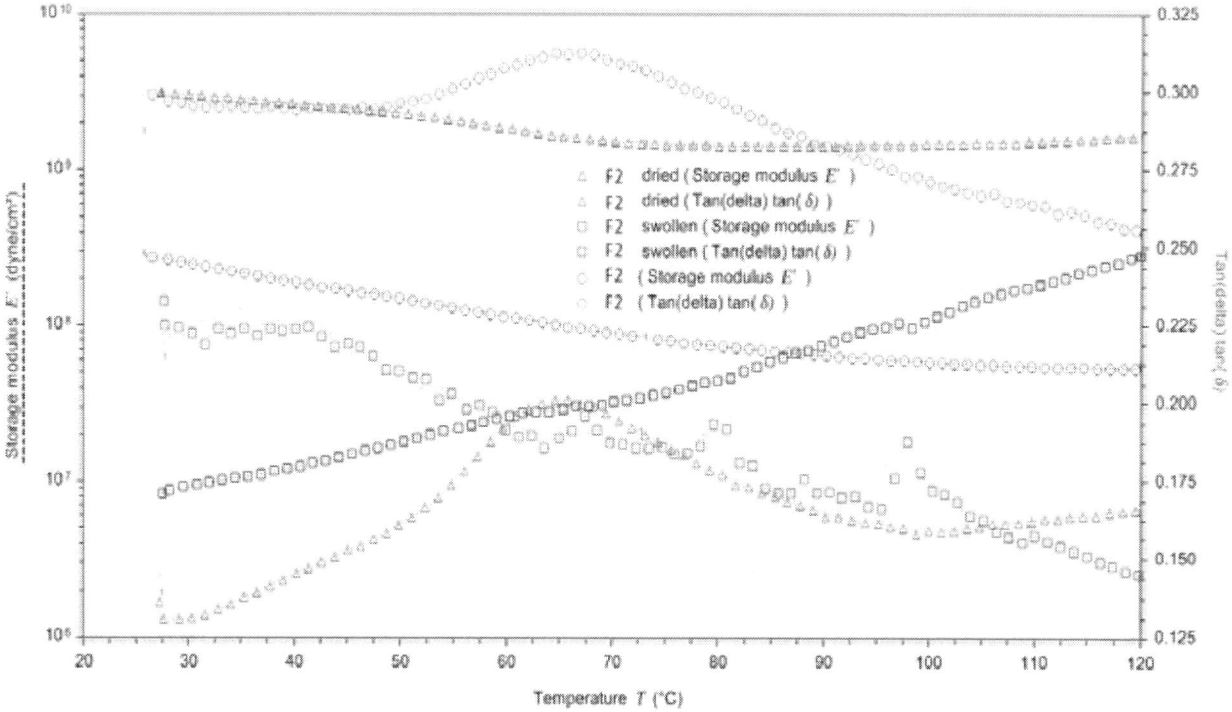

Figure 3. DMA data on F2 sample

It is important to mention that sample F2 exhibited a certain degree of brittleness after drying, which is undesired in rubbers, but this can be managed by incorporation of common ionomer plasticizers.

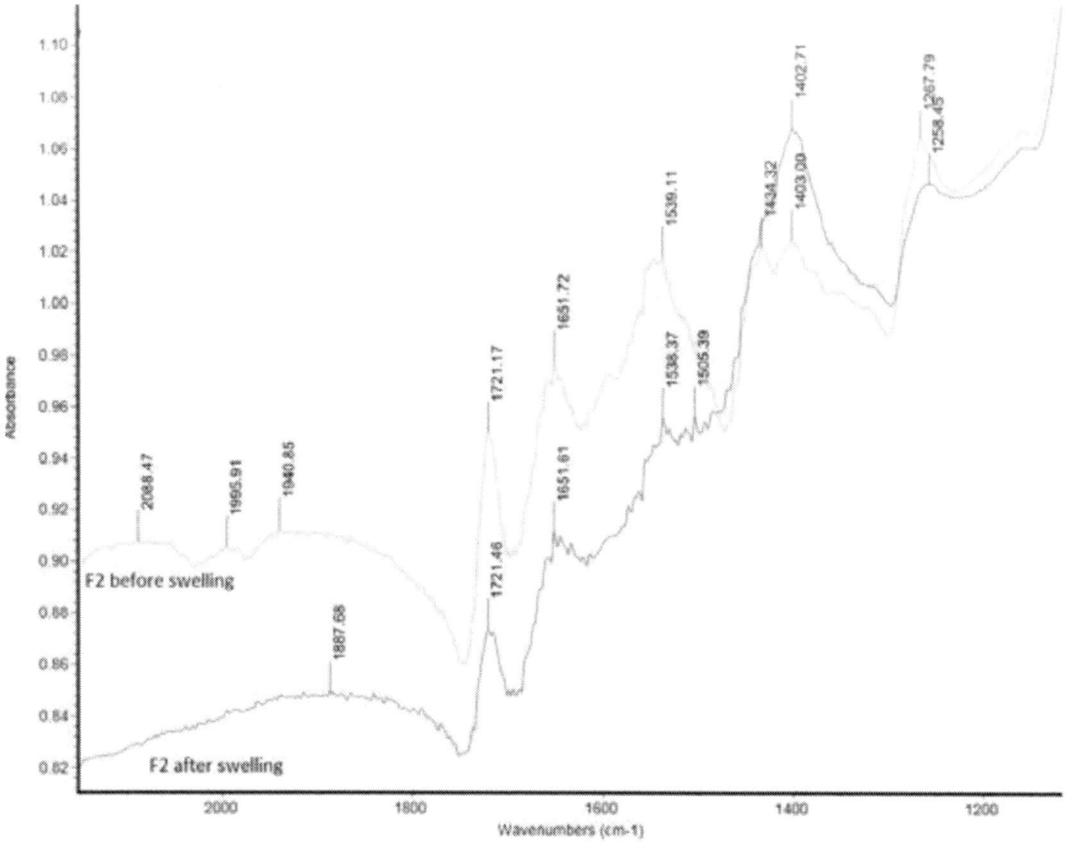

Figure 4. FTIR spectra of F2 sample taken before and after swelling

Conclusions

Fly ash demonstrated a great potential for consideration as cost effective, low reinforcing filler for downhole oil and gas applications. Although the reinforcing properties of fly ash are inferior to carbon black, incorporation of fly ash can improve the mechanical properties of HNBR elastomer compared to the unfilled raw crosslinked elastomer at both room temperature and at 325°F.

Furthermore, composites of HNBR with fly ash Class C demonstrated a stimuli-responsive effect by significantly increasing their modulus upon swelling in water brine. This effect was even more pronounced in composites of fly ash with swellable water compounds, which provided an easy access of filler to water upon swelling. A ten-fold increase in modulus was achieved in such fly ash reinforced sample, after swelling in 3% KCl water brine and drying in the oven overnight. The FTIR confirmed neutralization of carboxylic groups present in the investigated system, which contributed to overall improvement of mechanical properties of the compound upon swelling.

444310 Formal Modeling of Accidents and Automatic Generation of Tutoring Scenarios for Intelligent Safety Training of Process Operators

Abstract:

With the emphasis of accident-free operations and prevention of catastrophic events, safety-training demands for the chemical plants have been on the increase globally. However, the research results on modeling of accidents and involving human operators and the advanced theory and technology for (safety) operator training have been shared limitedly among domain experts and commercial developers. Even with the recent trend of simultaneously delivering the operator training system with the construction of a new process plant, in commercial operator training systems (OTS), safety training used to implement quite limited and obvious scenarios (and the burden was almost up to the trainer), partially due to the difficulty in modeling propagations of probable accidents. By the regulatory requirements or proactive investment by companies, effective scenarios for training of safety and abnormal situation management are necessary to be improved in the aspects of the size of volume as well as the training depth. Thus, considering the emergence of enhanced virtual reality technologies and inexpensive gadgets like HMD, supporting the immersive training environment, the future efficiency of the training heavily relies on the quantity and quality of safety scenarios, which would be essential in guiding the entire training process.

In this research we first overview a government-sponsored research project developing an enhanced safety training system for chemical process operators, using haptic devices in addition to enhanced audio and visual equipment. The initial version of the system runs with a fixed set of training scenarios developed for predetermined accident scenarios. However, with a limited set of training scenarios, the trainees simply memorize and rote learn the steps, required to be mastered to proactively solve the problems with an advent of possible accidents. In addition to that, without formal modeling, the system cannot give timely feedbacks and the performance of the trainee cannot be measured in real-time. Then to enhance the contents and usefulness of the training, the knowledge of process operations is collected and systematically stored as a form of knowledge base, in addition to the accident scenarios obtained by executing Fault Tree Analysis and HAZOP study.

By using a formal modeling language like UML and Petri nets, the accident scenarios and operators' problem-solving tasks are modeled hierarchically and classified into distinguishable subtasks. Then a new training scenario is generated on-line through automatic synthesis based on planning using the knowledge base. Safety training scenarios are generated intelligently for the tutoring of a single operator's job function as well as effective collaboration among a group of operators for real-scale malfunction diagnosis and intervention at the initial propagation stage of an accident. As case study, safety trainings for city gas governing stations are executed based on systematic scenario generation and analyzed for improved effectiveness and field applicability of the system. Reduction of time and cost in the training scenario production and real-time customized training based on new operator's level of knowledge and skill are also discussed in the framework of intelligent tutoring systems.

Copyright © American Institute of Chemical Engineers. All rights reserved.

444376 Ozone Source Apportionments of Local Anthropogenic Emissions

Abstract:

Ozone is one of six principal air pollutants designed by US EPA. NOX and VOC emissions from anthropogenic sources are the most important factors to control ozone pollutant.

In this report, photochemical dispersion models are used to estimate the effect of ozone from local anthropogenic emission sources. EKMA-type diagrams are generated for NOX- and VOC- effects. Furthermore, APCA (Anthropogenic Precursor Culpability Assessment) technology is used to estimate ozone source apportionments.

EKMA-type diagrams show that 10-30% emission reductions could reduce ozone concentrations to meet standard. Further APCA results show the average ozone contributions from regions and emission categories. APCA results explain different effects for different monitoring stations.

Copyright © American Institute of Chemical Engineers. All rights reserved.

444397 Flare Minimization and Emission Characterization during Start-Ups of Chilling Train in Ethylene Plants

Abstract:

Flaring is a crucial safety operation to protect chemical plant personnel and equipment during plant start-up. However, it generates huge amounts of off-spec products, and causes negative environmental and social impacts, as well as results in tremendous raw material and energy losses which could be saved to generate much more needed products. Thus, cost-effective start-up flare minimization strategies which is a double-win practice, simultaneously benefiting both the industrial and environmental sustainability, are becoming more and more important and attractive to the industries.

The chilling train is a very important section for the ethylene plants and accounts for more than 50% of the total start-up time and flare emissions of the whole ethylene plants. Within this section, the charge gas from the upper stream process is refrigerated to a very low temperature (lower than -160 °C) to separate hydrogen and methane. Meanwhile, the left charge gas is liquidized for cryogenic separation. The chilling train system consumes over 40% of the refrigerating capacity of an ethylene plant. During the chilling train start-up, different operational conditions with different compositions and flowrate of precooling process will affect both product separation efficiencies, start-up stability time and flare emission.

Traditional start-up scenarios are developed based on experience and improved by trial and error, which is inefficient and dangerous. In this study, a general methodology on flare minimization for chilling train start-up operations via rigorous dynamic simulation is presented. The methodology starts with setup and validation of chilling train steady-state and dynamic simulation models. The initial state of start-up in the real plant is achieved through systematically transforming from the validated dynamic model, and several operation conditions with different compositions and flowrate of precooling stream are examined to identify an optimal initial state of start-up based on the obtained initial state. Thereafter different flare minimization strategies for the chilling train start-up are conducted on the optimal initial state model. Also, emission source characterization under different flare minimization strategies for chilling train start-up is investigated. Deep insights of the start-up strategies, emission source distribution and dynamic emission profiles are provided. This study provides detailed technical support for both the industry and environmental agencies on evaluating and developing cost-effective flare minimization strategies in the future.

Copyright © American Institute of Chemical Engineers. All rights reserved.

Experimental Study on Najas Minor of Producing Bio-Oil

Abstract: Due to the diminishing of petroleum reserves and the environmental degradation caused by consumption of fossil energy , many attentions have been paid to the application of biomass energy which is acceptable as an abundant renewable energy resource. Najas minor grows in dense clusters with highly branched stems of which a small fragment could propagate into a thrive one.In this study, we researched the properties of Najas minor to make biofuels .Firstly,the najas minor after heavy metal ions degradation were fed into pyrolyser to prepare biofuels. Significant differences of the product was observed compared with only Najas minor pyrolysed.

KEY WORDS: Najas Minor; catalytic pyrolysis; bio-oil

1.Introduction

Energy demands and environmental problems caused by fossil energy have been a matter of growing concern.Bio-oil has been considered to be a promising way for producing renewable bio-fuels or chemicals[1-3].In recent years,many countries pay attention to the importance of biomass that is an abundant renewable energy resource.Biomass can be defined as any organic material mainly consisting of carbon, hydrogen,oxygen, nitrogen, and some other components in small proportions[4].Although the biomass has been studied for nearly a century,most studies mainly concentrated in Lemna minor, Lemna gibba, Spirodela polyrhiza and Landoltia punctata.We still know little about the biological characteristics and potential application value of other varieties of biomass[5].Najas minor,a submersed herb,grows in dense clusters with highly branched stems of which a small fragment could propagate into a thrive one.Najas Minor is native to Europe, Asia and North Africa from the Netherlands to Morocco east to Japan and the Philippines,including China,Siberia, Central Asia, Iran, Turkey, Ukraine, Germany, France Italy and a host of other Countries.

Catalysts used for catalytic cracking are basically divided into the following categories: molecular sieve, alkali metal salt, metal oxide, etc.Different catalysts have different catalytic effects.The use of metal nanoparticles has shown to be very important in recent industrial applications[6].The application of precious metal nanometer materials in catalysis,which has caused the extensive concern of researchers in recent 30 years,has become a hot research topic in the field of catalysis.Gold nanoparticles have received a great deal of attentions due to their special optical and electrical performances, catalytic properties and remarkable quantum size effect, surface effect etc[7]. The synthesis of metal nanoparticles can be classified broadly into wet and dry methods[8]. The cost of production is relatively high both materially and environmentally. There is an unequivocal need to develop more cost-effective and environmental benign methods to alternate these existing methods[9].Researchers in this field turn to use organisms to produce inorganic nanostructures extracellularly[10].

Although there has been a large volume of work on the synthesis of gold nanoparticles[11-13], and obtainning bi-oil by abundant biomass, there are relatively fewer attempts to combine these two methods.In this work, gold nanoparticles were obtained by aqueous chloroauric acid solution with the of Najas Minor at room temperature. The direct pyrolysis and catalytic pyrolysis experiments of Najas Minor were carried out to observe the effects of catalysts on pyrolysis process of Najas Minor.

2. Materials and Methods

2.1. Experimental samples

The Najas Minor collected from Tianjin city and the solutions of Au(III) were used in this study.The stock solutions of Au(III) were prepared by dissolving analytical grade chloroauric acid

($AuCl_3 \cdot HCl \cdot 3H_2O$, Kewei Reagent, China) in deionized water.

2.2.Experimental apparatus and procedure

Pyrolyzer with GCMS,shown in Fig. 1,is mainly used for analyzing the composition of polymer.Pyrolyzer with GCMS includes three parts :quartz pyrolysis tube,pyrolyzer and GCMS.

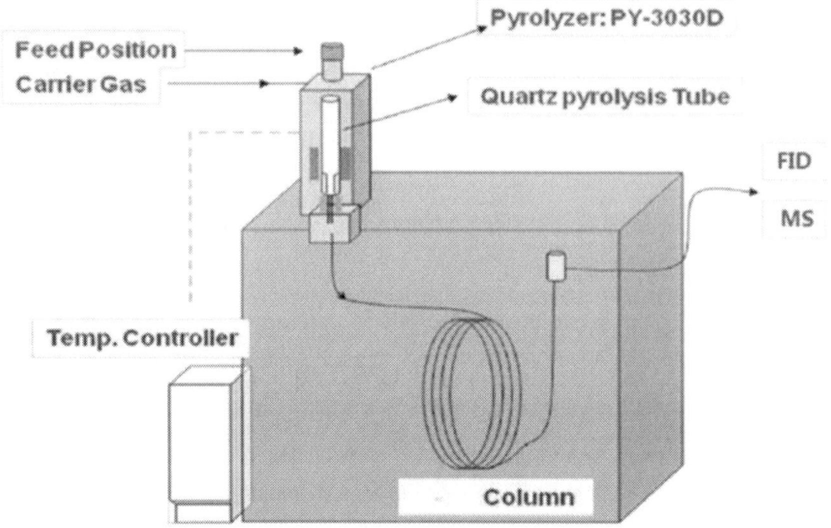

Fig. 1 Diagram of Pyrolyzer with GCMS

The quartz pyrolysis tube (shown in Fig.2) length is 12cm.The sample end diameter is 5 mm,and the pyrolysis products end diameter is 3mm.This special structure is beneficial to the fixation of the reactant in the tube.The top and bottom of the reactants are both supported by quartz wool to prevent reactants blowed by the carrier gas ,or fine particles into the chromatographic column.All pyrolysis experiments were carried out on the EGA/PY-3030D pyrolyzer(Frontier Lab,Japan) .The samples were filled in the special quartz pyrolysis tube to reduce the amount of valuable catalyst and reduce the cost of research work.

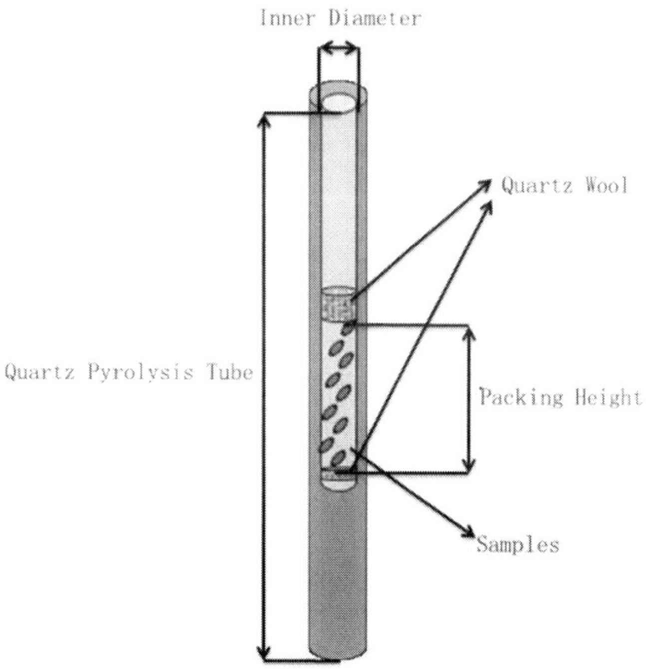

Fig. 2 Diagram of Quartz Pyrolysis Tube

Approximately 0.5g Najas Minors were used in each experiment.The samples were impregnated in 50ml deionized water and 50ml solution of chloroauric acid(0.0793mmol/L) for a night,followed by drying at 80℃ until the weight didn't change.Each experiment used 2.5mg Najas Minors as reactants.Each sample was heated, under the control of heating controller, from ambient temperature to 600 ℃ repeatly and kept at 600 ℃ for 2 h.

3. Results and discussion

3.1.MS Analysis of the Products during the Pyrolysis

Under the same pyrolysis conditions, bio-oil obtained by direct pyrolysis and catalytic pyrolysis were analyzed by GCMS,and the main chemical compounds were shown in Table1 and Table2.

Table 1 Indentification and quantification of main chemical compounds by direct pyrolysis

NO.	peak area%	Formula	Compound
1	12.67	$C_3H_6O_2$	2-Propanone, 1-hydroxy-
2	3.67	C_7H_8	Toluene
3	3.28	C_7H_8O	p-Cresol
4	3.81	$C_5H_{10}O$	Pentanal
5	3.53	C_8H_8O	Benzofuran, 2,3-dihydro-
6	2.17	C_8H_8N	Indole
7	5.93	$C_{20}H_{40}O$	3,7,11,15-Tetramethyl-2-hexadecen-1-ol
8	2.52	$C_{22}H_{42}O_2$	Phytol, acetate
9	5.25	$C_{16}H_{32}O_2$	n-Hexadecanoic acid
10	8.99	$C_{18}H_{30}O_2$	9,12,15-Octadecatrienoic acid, (Z,Z,Z)-

Table 2Indentification and quantification of main chemical compounds by catalytic pyrolysis

NO.	peak area%	Formula	Compound
1	6.01	C_7H_8O	p-Cresol
2	1.75	$C_7H_7NO_2$	3-Hydroxypyridine monoacetate
3	4.58	C_8H_7N	Indole
4	3.05	$C_{10}H9N$	1-Phenyl-1-cyclopropanecarbonitrile
5	3.90	$C_{12}H_{22}O_{11}$	.alpha.-D-Glucopyranose, 4-O-.beta.-D-galactopyranosyl-
6	8.04	$C_{20}H_{40}0$	3,7,11,15-Tetramethyl-2-hexadecen-1-ol
7	2.82	$C_{20}H_{40}$	2-Hexadecene, 3,7,11,15-tetramethyl-, [R-[R*,R*-(E)]]-
8	3.99	$C_{15}H_{28}$	1-Pentadecyne
9	2.12	$C_{15}H_{28}$	3,7,11-Trimethyl-2,4-dodecadiene
10	8.75	$C_{16}H_{32}\,0_2$	n-Hexadecanoic acid

3.2 discussing

According to the chemical category and common characteristics, the products were divided into hydrocarbons, alcohols, acids, ketones, nitrogen compounds and other compounds.Nitrogen compounds detected by GC/MS analysis ,mainly come from the pyrolysis of amino acids in Najas Minor.Table 3 showed the distribution of pyrolysis products.As shown in the table,the hydrocarbons ,alcohols and nitrogen compounds in biological oil of catalytic pyrolysis were significantly increased compared with direct pyrolysis.The content of hydrocarbons increased from 12.4% to 34.6%.The content of alcohols increased from 9.1% to 16.6%.The gold nanoparticles loaded in surface of Najas Minor, was conducive to the generation of hydrocarbon and alcohol in pyrolysis.By analyzing the GC/MS results, the main hydrocarbons in bio-oil by catalytic pyrolysis were long chain alkanes,and olefins.The main components in gasoline and diesel is hydrocarbons. As a result, whether the biomass can be effective as future power fuel oil, depends largely on the hydrocarbon content of bio-oil after further treatment.

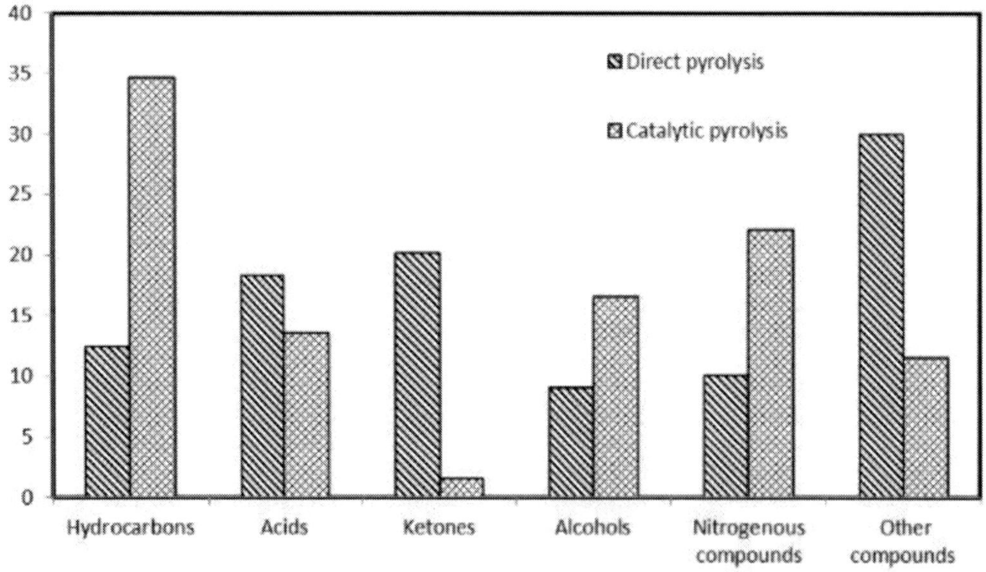

Fig.3 Distribution of pyrolysis products

4.Conclusion

Biomass was explored as a potential biomass energy resource. The experimental study on the direct and catalytic pyrolysis of Najas Minor had been presented, using Pyrolyzer with GCMS,to observe the catalytic effects of gold nanoparticles.In this study,remarkable catalytic effects for pyrolysis were observed.The pyrolysis of Najas Minor with gold nanoparticles,provided more hydrocarbons, alcohols and nitrogen compounds.

参考文献:

[1]Gong F, Yang Z, Hong C, et al. Selective conversion of bio-oil to light olefins: Controlling catalytic cracking for maximum olefins[J]. Bioresource Technology, 2011, 102(19):9247-54.

[2]Zhang Z, Bi P, Jiang P, et al. Production of gasoline fraction from bio-oil under atmospheric conditions by an integrated catalytic transformation process[J]. Energy, 2015, 90:1922-1930.

[3]Hui S, Yuqi D, Jinzhao D, et al. Transesterification of sunflower oil to biodiesel on ZrO2 supported La2O3 catalyst.[J]. Bioresource Technology, 2010, 101(3):953–958.

[4]Yaman S. Pyrolysis of biomass to produce fuels and chemical feedstocks[J]. Cheminform, 2004, 35(31):651-671.

[5]Hejný S. Ellias Landolt The family of Lemnaceae—A monographic study—Vol.1 biosystematic investigations in the family of Duckweeds (Lemnaceae) (vol. 2)[J]. Folia Geobotanica Et Phytotaxonomica, 1993, 28(1):50-50.

[6]Gardea-Torresdey J L, Tiemann K J, Gamez G, et al. Gold Nanoparticles Obtained by Bio-precipitation from Gold(III) Solutions[J]. Journal of Nanoparticle Research, 1999, 1(3):397-404.

[7]Marie-Christine D, Didier A. Gold nanoparticles: assembly, supramolecular chemistry, quantum-size-related properties, and applications toward biology, catalysis, and nanotechnology.[J]. Chemical Reviews, 2004, 35(16):293-346.

[8]Liu B ,, Xie J ,, Lee J Y, et al. Optimization of high-yield biological synthesis of single-crystalline gold nanoplates.[J]. Journal of Physical Chemistry B, 2005, 109(45):15256-63.

[9]Raveendran P, Fu J, Wallen S L. Completely "Green" Synthesis and Stabilization of Metal Nanoparticles[J]. Journal of the American Chemical Society Jacs, 2003, 125(46):págs. 13940-13941.

[10]Absar Ahmad , Satyajyoti Senapati , MIK, et al. Extracellular Biosynthesis of Monodisperse Gold Nanoparticles by a Novel Extremophilic Actinomycete, Thermomonospora sp.[J]. Langmuir, 2003, 19(8):3550-3553.

[11]Selvakannan P, Mandal S, Phadtare S, et al. Water-dispersible tryptophan-protected gold nanoparticles prepared by the spontaneous reduction of aqueous chloroaurate ions by the amino acid[J]. Journal of Colloid & Interface Science, 2004, 269(1):97-102.

[12]Gittins D I, Caruso F. Biological and Physical Applications of Water-Based Metal Nanoparticles Synthesised in Organic Solution[J]. Chemphyschem, 2002, 3(1):110－113.

[13]And A G, Murphy C J. Seed-Mediated Synthesis of Gold Nanorods: Role of the Size and Nature of the Seed[J]. Chemistry of Materials, 2004, 16(19):3633-3640.

444353 A Combined Rheology and Stability Study on Emulsified Acid for High Temperature Applications

Abstract:

HCl emulsion has been a popular formulation to stimulate carbonate reservoirs due to the various advantages that emulsified HCl has over ordinary HCl; including but not limited to deeper penetration and less corrosion damages. Emulsified HCl has even more potential advantages if utilized properly. The reaction kinetics, stability and rheology of emulsified HCl plays a significant role on achieving successful acidizing treatment.

The objective of this paper is to experimentally study the factors affecting the stability of emulsified HCl. The emulsion was prepared using 15wt% HCl and diesel with cationic emulsifier. The acid solution-to-diesel ratio was 70:30. The effect of stirring intensity, concentration of the emulsifier, addition of corrosion inhibitor, temperature, addition rate of acid phase to oil phase and mixing time of the emulsion have been studied. Stirring intensity was varied from 600 to 1500 rpm. Emulsifier dosage ranged between 5 to 20 gpt. Temperature used for heating, to mimic reservoir condition, was as high as 120°C.

The results showed that there is an optimum condition for emulsifier dosage and for mixing time of the emulsion. The optimum emulsifier dosage was found to be 10 gpt with 5 minutes of mixing time. As for the addition rate of acid phase to oil phase, 1.4 ml/min of drop-wise addition was found to give a very stable emulsion. The effect of stirring intensity on the emulsion stability was proportional, the higher the rpm speed the better the stability. Corrosion inhibitor, on the other hand, played a role as a destabilizer. With regards to heating temperature, it was found out that the higher the temperature, the faster the emulsion breaks.

The novelty of this study is the comprehensive approach in combining stablity and rheology of emulsified acid for high temperature applications. The particle size analysis and solution morphology help in better understanding HCl emulsion and how it works.

Copyright © American Institute of Chemical Engineers. All rights reserved.

COMPARATIVE ANALYSIS OF MOLECULAR STRUCTURE IDENTIFIABILTY BASED ON SIGNATURES AND DESCRIPTORS

Zelimir F. Kurtanjek
University of Zagreb, Croatia

Abstract

This work is focused on molecule computer representation aimed for molecule predictive model (QSAR) and inverse modelling. Two different approaches are compared, i.e. the molecule structure coding based on graph theory (Faulon et al.[1] extended valence) and the chemical based descriptors[2]. Applied are 1-3rd level molecule graph fragments and 630 2D molecular descriptors. The molecule test sets are alkanes, alkenes, acetones, aromatics, organic acids and halogenated hydrocarbons in the range of C_1-C_{12}, selection of biomolecules (including aminoacids) and a set of binary ionic liquids (cations: imidazole, pyridinium, quinolinium, ammonium, phosphonium). The comparative comparison of molecule description based prediction potential are statistically evaluated by octanol-water (PWA) partition coefficients (for hydrocarbons), viscosity of ionic liquids and biological impact on acetycholinesterase inhibition. The evaluation is based on two distinct levels. On the first level the comparative analysis based on internal and external correlation between "scores" in multidimensional spaces of the separate first, second and third order valance spaces, and the molecular signature space. Firstly, the potential of information content applicable for inverse modeling is tested on ability for molecule classification by use of linear discrimination (LDA) based on the singular value decomposition (SVD) of corresponding molecular fragments and descriptors. For chemical-biological property inverse modelling proposed is application of constrained genetic algorithm (GA). Tested are linear structure multivariate (chemometric) models such as principal component regression and partial least squares and nonlinear decision trees (random forest). Applied are singular value decompositions for reduction of the molecule signature and the descriptor spaces up to the significance level of p=0.05. The selected level of significance is presumed to be in coherence with experimental data of molecule physical and chemical properties. The main focus is on extraction of the main molecular fragments (partial graphs extracted from molecular signatures) and specific molecule descriptors. Importance of application of open source software modules provided by Bioclipse[3] and R software packeges[4] for computational-statistical numerical algorithms is emphasized.

Key words: Molecule signatures, descriptors, inverse modelling, decision tree

Introduction

Advancement of high-throughput experiments have lead to „big data" databases applied in pharmaceuticals, medicine, new material science, catalysis, and new industrial processes. The role of chemoinformatics as a science for computer encoding of molecule properties and algorithms for data mining and inference are the essential steps in the development. Computer encoding of molecular properties has been a part of chemical engineering from the beginning of process design. It is mostly based on molecule group contributions, but has been extended by application of molecular descriptors (1-2-3 D), molecule signatures, and application of molecular quantum mechanics. Modeling is faced with the problem of „high dimension curse". For example, about 5 000 various molecular descriptors can be determined for each molecule, although most of them are highly correlated. It results in an ill-conditioned problem of modelling with large uncertainties in parameter and poor model prediction accuracy outside modelling training sets. Great improvement has been achieved by application by use of model ensembles, like random forest of many individual decision trees. These models are highly flexible, can be trained on discontinuous and nonlinear functional relations, can encompass various data formats like numerical and categorical variables, and due to large number of randomly selected individual predictors in principle do not have tendency to overfitting. However, when the object of modelling is extraction of important molecular features, due to autocorrelation between descriptors random forest, produced lists of importance are large sets of poorly distinguishable variables. The focus of this work is to apply feature regularization for optimization and reduction of list of variable importance. Compared are two methods for structure encoding, molecular descriptors and molecular signatures of first and second order.

Modeling and Factor Regularization

In this work are selected two distinct data sets of organic molecules from open source data bases (J. Sangster, 1989) and ionic liquids ILs (The UFT/ Merck and NIST, 2015), and lists of the selected molecules are given in Table 1. As target properties for organic molecules are selected the corresponding molecular classes and partition octanol-water (POW) coefficients, and dynamic viscosity in the range of 30-80 ^{0}C of ILs is selected The molecules are coded into SMILES and used as input lists into software for evaluation of molecular signatures. Most popular open source software tools are DRAGON (R. Todeschini, et al., 2009) and PADEL (Chun Wei Yap, 2010). Here are used 1-2 D descriptors generated by PADEL. For the same list of molecules are determined first and second order molecular signatures or extended valances (J.P. Faulon, 2003). The signatures are two dimensional graph based structural molecular fragments. The vertices are atoms and the edges are corresponding bonds (lengths or heights), and are here only used of the first and second level. A molecule is characterized by a complete set of signatures as input data for modelling. Since the models are based on large number of variables (molecular features) possibly lager than number of samples (molecules), usually denoted as p>>n models. They are intrinsically „ill conditioned" leading to good predictions on train data but very poor predictions on test data. Also, produced list of variable importance are very dense "clusters are not distinguished" with large uncertainties (broad confidence intervals). This makes the problem of inverse modelling aimed to predict a set of candidate molecules for given set of target properties, such as in a case of high dimension combinatorial task for selection of ILs cations and anions, very unreliable. In order to reduce model dimensionality and improve robustness and parsimonious of important molecular features here is applied modelling with feature regularization. The model is based on ensemble of random trees (CART models, L. Breiman et al. 1998) with included feature regularization.

Data sets of organic molecules	Range of number of atoms				
Molecular classes	C	H	O	Cl	F – I - Br
Alkanes	5-12	12-30			
Alkenes	4-9	6-18			
Acetones	6-24	6-26			
Halogenated hydrocarbons	3-10	6-22	1		
Organic acids	1-16	2-32	2		
Esters	4-11	6-22	2		
Aromatics	1-12	1-21		3	2 – 2 - 1
Data sets of ionic liquids (ILs)					
Cations	**Anions**				
1,3-Dimethylimidazolium (+)	PF_6 (-) Cl (-) CH_3COO (-)				
1-Methylpyridinium (+)	BF_4 (-) $MeSO_4$ (-) $EtSO_4$ (-)				
1,1-Dimethylpyrrolidinium (+)	Tf_2N (-) CF_3SO_3 (-)				

Table 1. The molecular groups of the organic and ionic liquid molecules selected for random forest models and extraction of the most important molecular features. The properties for the organics are obtained from J. Sangster (1989) and for the ionic liquids from Merck and NIST data bases (2015). Molecular descriptors and signatures of ILs ions are recorded separately and adjoined as a single record.

The modelling is formulated as an optimization (functional and multivariable minimization) problem with two parts of the objective function (T. Hastie et al., 2009).

$$\min F_{object.}(X, T, \omega) = SSE(y, \hat{y}) + \phi(T, \omega) \tag{1}$$

The first part is the sum of square errors (SSE) based on difference between observed y and prediction $\hat{y}$ based on input data by an ensemble T of interrelated CART models (L. Breiman et al. 1998), The second part is a selected metrics ϕ for the model complexity depending the sum of leaf probabilities for population of individual CART models and their total number N_T .

$$\phi(T, \omega) = \gamma \times N_T + \frac{1}{2} \times \lambda \times \sum_i^{N_T} \omega_i^2 \tag{2}$$

The ensemble output prediction is a sum of predictions of individual T_i tree CART models:

$$\hat{y} = \sum_i^{N_T} T_i(X) \tag{3}$$

A very efficient open source code Extreme Gradient Boosting XGB (T. Chen, 2016) is available in R language (R Developm-ent Core Team, v. 3.2.3, 2015). The optimization algorithm is consecutive, starting with first CART model which is appended with a next tree boosted by errors in predictions of previous tree. At each step optimization is solved by linearization of the objective function to quadratic terms. Number and structure of trees as well as the parameters for regularization need to be found by trial and error and minimal number of trees by folding and cross validation error on the training set. The critical step is validation on untrained set to have acceptable model error and avoid model overfitting with minimal model complexity.

Results and Discussion

Tested are models for prediction of the seven molecular classes (Table 1.) based on molecular descriptors and molecular signatures (first and second level). Results for molecular descriptor feature importance are shown on Fig. 1. The effect of regularization is very pronounced. Out of hundreds of descriptors for the molecular classification needed are only few, the most important being BCUTw.1h. On contrary, when an ensemble of independent CART trees is applied, i.e. random forest model, level of accuracy of prediction is the same, but the list of variable importance is densely populated with numerous descriptors.

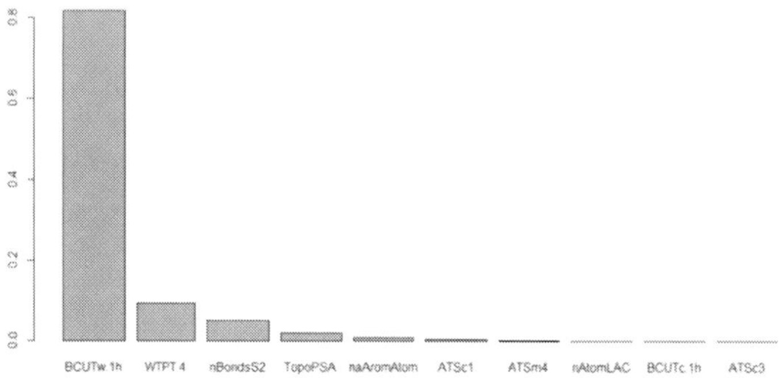

Fig 1. Variable importance molecular descriptor factors for the inverse modeling of the selected molecular classes. The first five descriptors are: BCUTw.1h (based on a weighted version of the Burden matrix which takes into account both the connectivity as well as atomic properties of a molecule), WTPT.4 (Randić's weighted path), nBondS2 (bond count), and naAromAtom (number of aromatic atoms).

The same classification problem was tested by the first and second order molecular signatures. The important signatures are depicted on Fig. 2-3. Accuracy of prediction on trained sets of molecules compared to descriptors is about the same, though the importance factors are not as distinctive as in case of the descriptors.

significantly better when molecular descriptors are applied. The

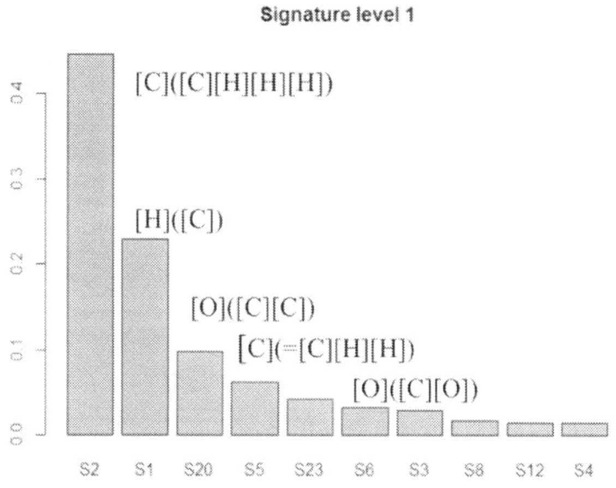

Fig 2. Relative importance of the individual first level signatures for inverse prediction of the selected molecular classes. The first five most important signatures are coded according to SMILES (Simplified Molecular Input Line Entry System).

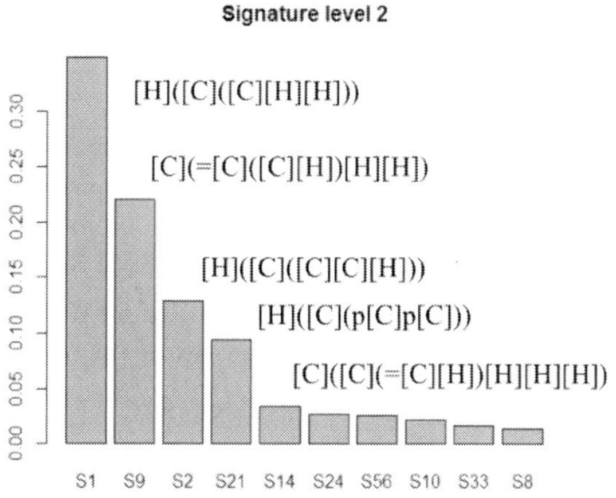

Fig 3. Relative importance of the individual second level signatures for inverse prediction for the selected molecular classes. The first five most important signatures are coded according to SMILES (Simplified Molecular Input Line Entry System).

However, model accuracy on untrained set of molecules, depicted in Fig. 4, is significantly lower for the model with molecular descriptors compared to signatures. The average error with descriptors is about 5% while for signatures about 11%. The "box and whiskers" plots are obtained by error evaluation of ten randomly generated untrained sets.

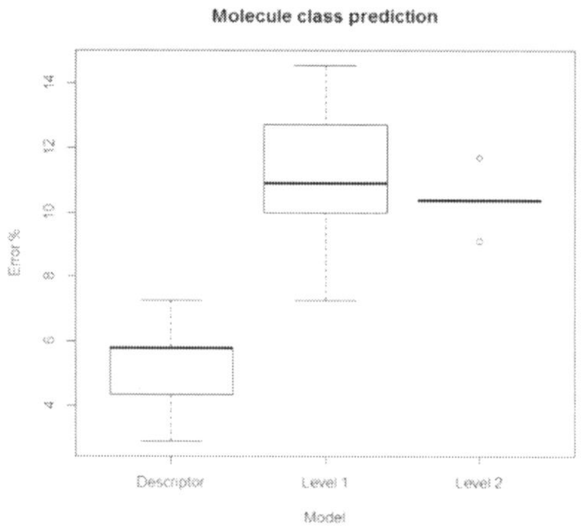

Fig 4. "Box and whiskers" plot for the model validation (relative errors) for inverse prediction of the molecular classes based on the set of untrained molecular descriptors and the first and second level of molecular signatures.

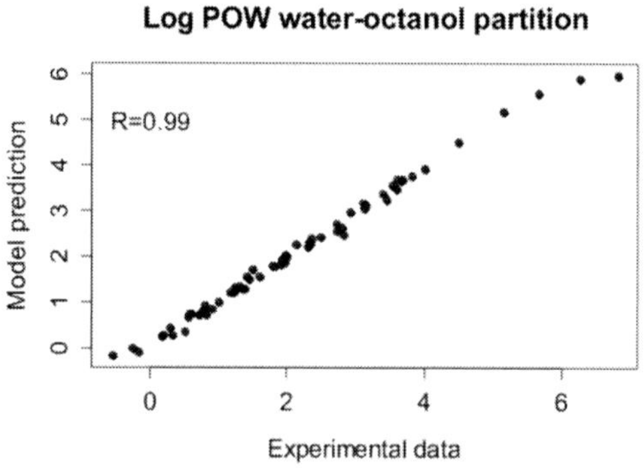

Fig 5. Predictions of the partition octanol-water coefficients POW, for the specified molecular classes of organic molecules, are based on the random forest model with the molecular descriptors.

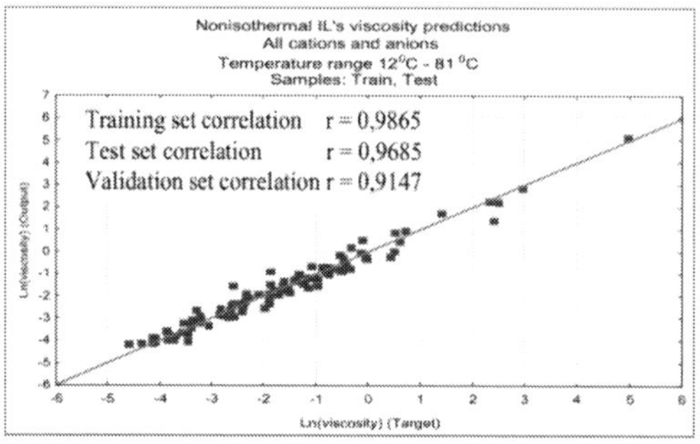

Fig 6. Predictions of dynamic viscosities for the specified groups of ionic liquids (ILs) based on the random forest model with the molecular descriptors. The viscosities cover the temperature range from 12 – 81 ^{0}C and five orders of magnitude.

In Fig. 4 can be observed that dispersion of errors (quantiles) for the second level molecular signatures is negligible, though two outliers are present.

For prediction of molecular properties applied are random forest models based on molecular descriptors. For all of the organic molecules, included in the seven classes, evaluated are octanol-water partition coefficients POW, shown in Fig. 5. The range of POW values is very high, on log scale fro 0 to 6, but the obtained accuracy of the model predictions is high (Pearson correlation R=0.99). The same approach is applied for prediction of ionic liquid viscosity data in temperature range from 12 to 81 ^{0}C. Since temperature effects on viscosity are intrinsically nonlinear, they represent a modelling challenge. However, the decision trees are very efficient for accounting nonlinearities, though nonlinearity is reflected in non-orthogonal decision planes, rendering trees with significantly increased number of branches. For the range of 6 order of magnitude the accuracy of prediction on trained set is R = 0.98 and on untrained set R = 0.91.

Conclusions

Compared are ensemble decision trees models for molecular feature class and property prediction based on extensive list of molecular descriptors and first and second level of extended valances or signatures.

The aim was to test effects of regularization on ensemble of CART decision tree models. Proven is effectiveness of the extreme gradient boosting algorithm in cases with extensive list of (molecular) features.

Obtained is increased accuracy on untrained sets of data, and more importantly, obtained are parsimonious lists for ranking of variable importance lists. This enables further significant reduction in development of robust and simple decision tree models.

The reduction in number of molecular features applied for decision tree models may greatly simplify the challenging problem of inverse modeling based on molecular descriptors and/or signatures.

References

1. J. Sangster, J. Phys. Chem. Ref. Data, **18**(3)(1989) 1111-1227
2. The UFT/ Merck Ionic Liquids Biological Effects Database, www.il-eco.uft.uni-bremen.de
3. NIST, Ionic Liquids Database Standard Reference Database #147, <ilthermo.bouder.-nist.gov /ILThermo/mainmenu.uix>
4. R. Todeschini and V.Consonni: "Molecular Descriptors for Chemoinformatics", (2 volumes), WILEY-VCH,Weinheim (Germany) 2009.
5. Chun Wei Yap, PaDEL-Descriptor: An Open Source Software to Calculate Molecular Descriptors and Fingerprints, J. Comp. Chem., (2010):**32**: 1466-1474.

6. J.-L. Faulon, D.P. Visco Jr., R.S. Pophale, The signature molecular descriptor. 1. Extended valence sequences vs. topological indices in QSAR and QSPR studies, J. Chem. Inf. Comput. Sci. **43** (2003) 707–720.
7. J.-L. Faulon, C.J. Churchwell, J.D.P. Visco, The signature molecular descriptor. 2. Enumerating molecules from their extended valence sequence, J. Chem. Inf. Comput. Sci. **43** (2003) 721–734.
8. L. Breiman, J. H. Friedman, R. A. Olshen, C.J. Stone, „Classification and Regression Trees", CRC, Boca Raton, 1998.
9. T. Hastie, R. Tibshirani, J. Friedman, The Elements of Statistical Learning, Data Mining, Inference and Prediction, Springer Verlag, Second Edition, New York, 2009.
10. T. Chen, T. He and M. Benesty (2016). xgboost: Extreme Gradient Boosting. R package version 0.4-3. https://CRAN.R-project.org/package=xgboost
11. R Development Core Team, , R: A language and environment for statistical computing. R, Vienna, Austria,www.R-project.org 2015.

444791 Dissolution and Precipitation of Minerals Under Different Alkali Concentrations in a Reactive CO_2 Batch System

Abstract:

The primary focus of this research is to study the dissolution and precipitation of minerals and salts from Berea sandstone and brine samples taken from the Great Salt Lake in a reactive CO_2 batch system. Various experiments were constructed to measure dissolution and precipitation reaction rates under different alkali concentrations. Changing the alkali concentrations of the Great Salt Lake brine samples produced changes in dissolution and precipitation of minerals from the Berea sandstone and the Great Salt Lake brine samples. Data from ICP-MS was used to quantify mineralogical changes in the samples and Micro-CT was used to measure the porosity change in the Berea sandstone samples. The results from these analytical instruments showed patterns of mineralogical dissolution and precipitation in the post-experimental fluid that correlate to the change in alkali concentrations.

Copyright © American Institute of Chemical Engineers. All rights reserved.

444825 Significance of Energy Intensity and GHG Emission Reduction Towards Sustainability of a Petrochemical Plant

Abstract:

Sustainability is an important pillar in today's petrochemical industry for improving process efficacy by reducing specific energy consumption, greenhouse gas (GHG) emissions, and wastewater. SABIC (Saudi Basic Industries Corporation) like other giant petrochemical companies has set stringent sustainability targets and investing huge resources to transform its plants from the present position to the first quartile in each category.

SABIC is the world largest producer of ethylene oxide and ethylene glycol. For each ton of glycol produced, 0.2 to 0.5 tons of CO_2 is emitted as GHG. The significance of sustainability "in general" while the methodology adopted to reduce energy intensity, carbon footprint, and wastewater "in particular", will be discussed thoroughly in our talk.

Copyright © American Institute of Chemical Engineers. All rights reserved.

444827 Syngas to Ethylene Glycol: A Technology Overview

Abstract:

SABIC is the world largest MEG producer petrochemical company and most of its MEG is being produced through the hydrolysis of EO. Ethylene oxide (EO) is a simplest form of epoxides and due to its molecular structure, acts as a versatile intermediate for a variety of chemicals such as; monoethylene glycol (MEG), di- and triethylene glycols, ethoxylates, ethanol amines, and glycol ethers. The growing demand of MEG owing to its downstream utilization in polyester industry has prompted the research of an efficient, cost effective, and environmentally benign EO production process.

The major feedstock for MEG are ethylene and syngas. The heavy dependence of MEG producers on ethylene and its scarcity in some parts of the world promoted the need of alternate feedstock for highly demanded MEG. The technological developments utilizing syngas as a feedstock will be discussed along with the commercial analysis and future growth in EO/EG business.

Copyright © American Institute of Chemical Engineers. All rights reserved.

444834 Stability and Rheology of Emulsified Acid in Waste Oil: Effect of Emulsifier Concentration, Mixing Speed, Temperature and Droplet Size

Abstract:

Acid in oil emulsion is used in the industry for stimulation purposes. In particular, HCl as the acid phase and diesel as the oil. Emulsifying HCl has numerous benefits over regular HCl. Perhaps the main reason is due to deeper penetration. Apart from that, less corrosion damages are caused since the external phase is hydrocarbon. Several studies showed the success of replacing diesel as an external phase with other hydrocarbon oil, such as crude oil and xylene.

This work utilizes the extra hydrocarbon left unused –or sometimes dumped- from refineries. This unused hydrocarbon is referred to as waste oil. First the chemical composition of waste oil was studied using HPLC, SARA fractionation as well as FTIR. This helped in selecting the proper emulsifier agent for the waste oil. The HCl-waste-oil emulsion was prepared using 15wt% HCl with a ratio of 70:30 of acid-to-oil. Extensive work was done to ensure reproducibility of data. This study was done to experimentally investigate the effects of distinct important variables on thermal stability and apparent viscosity of the HCl-waste-oil emulsion. These variables are; emulsifier concentration, mixing speed and the addition of corrosion inhibitor. All experiments were conducted at a high temperature 120°C. To assess the quality of the HCl-waste-oil emulsion, droplet size was measured as a function of temperature (up to 100 °C).

The results showed that the HCl-waste-oil emulsion is a shear-thinning fluid. Power-law model was applicable seamlessly to all of the apparent viscosity data for all of measured temperatures. The optimum conditions were found to be; 0.5vol% and 600 rpm for emulsifier concentration and mixing speed, respectively. Apart from that, the emulsions prepared with corrosion inhibitor were less stable compared to the ones without it. However, the addition of corrosion inhibitor tended to generate monodispersed emulsion but bigger size droplets for temperatures 25°C and 50°C and polydispersed droplets for higher temperatures. Droplet size measurements showed an average size of 1.47 to 3.09 μm. Overall, the lab results showed a promising potential for the HCl-waste-oil emulsion to be used in the field.

This work takes into account; reducing the budget of acidizing job and at the same time environmental concerns. It makes use of the unwanted hydrocarbon from refiners. And, it saves the environment by using the waste oil instead of dumping it.

Copyright © American Institute of Chemical Engineers. All rights reserved.

444994 Multiscale Simulation of Proteins Motion in the Microchannel

Abstract:

Protein adsorption is an important phenomenon, which plays a prominent role in chemical, industrial and medical areas. In this work, the dynamic behavior of the egg-white lysozyme has been investigated in a micro-channel with a multiscale modeling approach. The Molecular Dynamic simulation, free energy computation and Langevion Dynamics (LD) simulations have been combined to study lysozyme adsorption process on Au (111) surface. The gold surface has strong attraction to proteins. The protein-surface adsorption or desorption free energies is calculated. From the case studies of LD simulations, the surface has effects on the protein dynamic adsorption process. By the comparison of LD, MD and the experiments, the time step is concluded as an important factor to study protein adsorption behavior at the water-surface interface. Fine time step at the scale of 1 picosecond is necessary to provide enough accuracy to resolve proteins' dynamics trajectories.

Copyright © American Institute of Chemical Engineers. All rights reserved.

445004 Grand Challenges in Future Energy and Chemicals: Climate Change and Sustainability As Drivers for New Technology and Innovation

Abstract:

Global demand for energy will double by 2050, driven by increasing population and affluence especially in emerging economies. While fossil fuels will continue to underpin the energy supply for the near future, concern over impact on climate change continues to grow, driving interest in expanded use of renewables and development of carbon capture and sequestration or re-use solutions. In addition to CO_2, water use and air emissions are critical issues to be addressed in future and current energy supply, as is energy storage for effective use of renewables. This presentation will provide an overview of what industry is doing today to address these issues, including the factors and metrics used in decision making.

Copyright © American Institute of Chemical Engineers. All rights reserved.

445029 Electrospun Nanocomposite Fibers for the Removal of Heavy Metals in Water

Abstract:

Uniform electrospun nanofibers with small diameters were successfully prepared by systematically optimizing the electrospinning parameters including polymer concentration, salt loading, applied voltage, tip-to-collector distance, and flow rate. These parameters are shown to affect the fiber morphology and the addition of iron (III) nitrate salt was found to be able to reduce the fiber size. Due to the conductive nature of iron salt in polymer solution, even a relative lower voltage of 6 kV could generate a stable fluid jet and yield the nanofibers. The thermal properties of the composite fibers were investigated by thermal gravimetric analysis (TGA) and differential scanning calorimetry (DSC). Rheological studies were also conducted. The composite nanofibers were further carbonized to carbon nanofibers. Applications of the nanofibers in environment remediation have shown promise.

Copyright © American Institute of Chemical Engineers. All rights reserved.

445030 Heavy Metal Detection in Treated Water Via a Fabric Nanocomposite Sensor

Abstract:

We present the conception of an electrochemical sensor for on-site/portable water analysis. This flexible sensor was developed to quantify the amount of heavy metals (in real-time), alleviating the wait time, large sample size, possible contamination, and expensive analytical equipment associated with current procedures for water analysis. This sensor is composed of a multiwalled carbon nanotube (MWNT) functionalized polymer mat, for a flexible yet conductive sensor. The MWNTs are then functionalized with a cyclo-oligomeric calixarene, which is shown to selectively form a supramolecular complex with metal ions. Upon complex formation, the charge carriers are drawn away from the carbon layer and the detection is prevalent at appropriate levels. The optimization of MWNT and calixarene functionalization, as well as the selectivity and sensitivity of the sensor will be discussed.

Copyright © American Institute of Chemical Engineers. All rights reserved.

447119 Development R&D of Manoir Industries and Emphasis on Coating

Abstract:

The Senior VP Technical Support & Projects will present the development R&D of Manoir Industries and especially the coating (it is our lattest development) as well the services we can provide such as metallurgy expertise, advice, convector cleaning, site welding, welding supervision in order to increase the furnace productivity. So it is 60% R&D and 40% services. Manoir Industries manufactures cast components for the petrochemical market and develops high temperature alloys such as the Manaurites.

Copyright © American Institute of Chemical Engineers. All rights reserved.

444478 Biogas Reforming with Addition of Carbon Dioxide to Syngas over Ni-Based Bimodal Pore Catalyst

Abstract:

Methane dry reforming of (MDR) has been widely studied due to its important applications in producing of syngas ($CO + H_2$), removal of two greenhouse gases, and upgrading of biogas (mainly composed of CH_2 and CO_2) into value added chemicals. Compared to the noble metal catalysts, nickel-based catalysts are suitable for industrial scaled-up MDR due to their high activity and low cost. However, nickel-based catalysts have been encountering a significant challenge that is the active metal sintering and coke deposition, leading to the catalyst deactivation and then the poor stability. Nevertheless, it has been reported that the deactivation of nickel catalysts can be suppressed by adding promoters like strong Lewis bases with the enhancement of chemisorb CO_2, such as MgO and CaO, and like lanthanide elements with the capacity of oxygen storage and release, such as CeO_2 and La_2O_3. Al_2O_3 is a commonly applied catalyst carrier own to its good pore size dispersion, high specific surface area and high mechanic strength.

Herein, we report that the bimodal NiCeMgAl catalysts with different NiO-loading were synthesized by the refluxed co-precipitation method and were evaluated for MDR performance by using real biogas from anaerobic digestion of food wastes. The sample containing 15wt% NiO-loading (Ni_{15}CeMgAl) was found to be active enough at 750 °C with a high CH_4 conversion of around 90%. Ni_{15}CeMgAl catalysts kept their bimodal porosity after reduction and MDR process. The evolution of the Ni_{15}CeMgAl catalyst before and after MDR process was investigated by SEM, TEM, XRD, and TGA techniques. The correlation between the structure evolution and catalytic performance change was also discussed.

Copyright © American Institute of Chemical Engineers. All rights reserved.

444551 Continuously Adjustable, Molecular-Sieving "Gate" on 5A Zeolite for Distinguishing Small Organic Molecules By Size

Abstract:

Zeolites/molecular sieves with uniform, molecular-sized pores are important for many adsorption-based separation processes. Pore size gaps, however, exist in the current zeolite family. This leads to a great challenge of separating molecules with size differences at ~0.01 nm level. Here, we report a novel concept, pore misalignment, to form a continuously adjustable, molecular-sieving "gate" at the 5A zeolite pore entrance without sacrificing the internal capacity. Misalignment of the micropores of the alumina coating with the 5A zeolite pores was related with and facilely adjusted by the coating thickness. For the first time, organic molecules with sub-0.01 nm size differences were effectively distinguished via appropriate misalignment. This novel concept may have great potential to fill the pore size gaps of the zeolite family and realize size-selective adsorption separation.

Copyright © American Institute of Chemical Engineers. All rights reserved.

16AIChE
Spring Meeting, Houston, TX

The Principle of Intrinsic Reality
Solving the World's Woes through Chemical Engineering and Economics

Jed Moyano Bellen, SCPM, PMP
Hyundai Engineering & Construction Co. Ltd.
9 Temasek Blvd., #27-02 Suntec Tower 2, Singapore 038989
jed.bellen@hdec.co.kr

Copyright © 2016 by Jed M. Bellen. All rights reserved. The views and opinions expressed in this article are those of the author and do not necessarily reflect the official position of Hyundai Engineering and Construction Co. Ltd.

Prepared for Presentation at
American Institute of Chemical Engineers
2016 Spring Meeting
11th Global Congress on Process Safety
Houston, Texas
April 10-14, 2016

UNPUBLISHED

AIChE shall not be responsible for statements or opinions contained in papers or printed in its publications

The Principle of Intrinsic Reality Solving the World's Woes through Chemical Engineering and Economics

Jed Moyano Bellen, SCPM, PMP
Hyundai Engineering & Construction Co. Ltd.
9 Temasek Blvd., #27-02 Suntec Tower 2, Singapore 038989
jed.bellen@hdec.co.kr

Keywords: Principle of Intrinsic Reality, Chemical Engineering, Economics, Sustainability

Abstract

Human beings in profound rationality always seek for explanations of the phenomena around them. To understand these phenomena, human beings have developed systematized body of knowledge as a practical way of dealing with life's challenges. Among the systematized body of knowledge developed are chemical engineering and economics.

An avant-garde way of looking at each of these disciplines leads to the Principle of Intrinsic Reality. This principle relates the fundamental laws of the two disciplines to glean insights on the constraints imposed by conservation laws on economic systems at both the macro and micro levels.

It can be argued that if mass and energy in the universe are constant, then logically diminishing returns and marginal utility are bound to be observed. This logic is quite evident in a microorganism culture which follows the Monod growth model.

It can also be thought that conservation laws imply scarcity of resources. As such, it enjoins everyone to find optimal and efficient use of resources in the context of trade-offs, production decisions and sustainability. It even calls for the development of new technologies to produce more outputs from less inputs (i.e. shifting of production possibility frontier to the right). In this way, the Principle of Intrinsic Reality leads to the imperative for sustainable optimization.

Furthermore, it is deemed that the conservation laws can explain the phenomenon called "polarity of wealth." As mass and energy when harnessed create wealth, then the constancy of mass and energy implies boundary for wealth creation. In effect, wealth can neither be created nor destroyed but can only shift from one entity to another thereby creating polarity in wealth distribution. This assertion can be the key towards more equitable restructuring of economic systems to close income gap and thereby result to stable and resilient populace.

In conclusion, this paper envisions for a paradigm shift from traditional money-centric view of chemical engineering and economics to a synergistic convergence of disciplines which can powerfully tackle the world's woes and thereby effect sustainability.

1. The World in which We Live Today

The past few years saw unprecedented global challenges which have put into question existing economic models and scientific paradigms. Among these challenges are the dystopian global economy, climate change, environmental degradation and terrorism. The magnitude of these challenges has grown to a profound level that newer ways of thinking and perspectives become necessary. With that end in mind, this paper attempts to look at the fundamental laws of chemical engineering and economics in new ways to recognize our intrinsic reality from which these pressing challenges can be tackled at large.

2. The Principle of Intrinsic Reality

It is an observable fact that advancement in science increasingly blurs boundaries of various disciplines in tackling proactively the pressing global issues or problems. Within this purview, two seemingly diverging disciplines namely chemical engineering and economics can be taken together to glean insights on our intrinsic reality and what it means for our economic and ecological systems on a larger scale.

In the author's contention, intrinsic reality is grounded on the fundamental laws of chemical engineering and economics. These laws are the Conservation Laws (Chemical Engineering) and Law of Diminishing Returns (Economics). It can be argued that the Law of Diminishing Returns is observable because of the constraints imposed by the Conservation Laws. As such, this leads to the Principle of Intrinsic Reality which can be stated thus;

> *In any system, both at the macro- and micro-levels, the limiting factor for growth is the constancy of potential inputs.*

Potential inputs mean not the actual inputs to processes but the constant material or energy available which can be harnessed to create wealth. The reasoning or proof of this principle is presented in subsequent discussions.

2.1 *Proof Based on the Monod Growth Model*

The best illustration to understand the Principle of Intrinsic Reality is the Monod Growth Model. Monod in his 1949 seminal paper presented an empirical relation between exponential growth rate and concentration of essential nutrient. The data was fitted into hyperbolic equation:

$$R = R_K\left(\frac{C}{C_1 + C}\right)$$

In this equation, R is the growth rate during the exponential phase; R_K is the rate limit for increasing concentrations of C; C is the concentration of the nutrient; and C_1 is the concentration of nutrient at which the rate is half the maximum.

Figure 1: Graphical Representation of Monod Growth Model

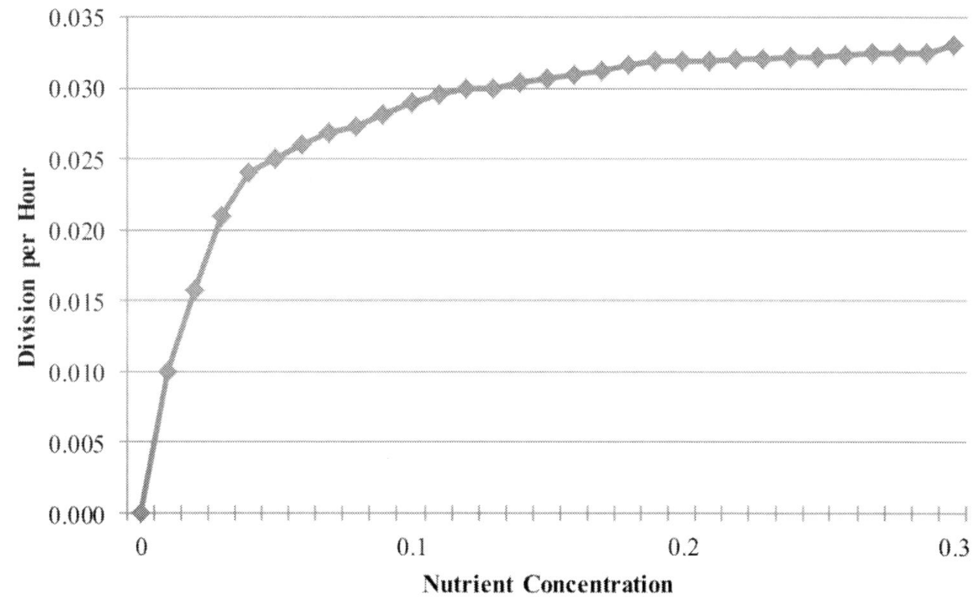

Monod (1949) further pointed out that the limiting factors commonly found are exhaustion of nutrients, accumulation of toxic metabolic products and changes in ion equilibrium.

The foregoing discussion has been observed at the micro-level with the bacterial culture. At the macro-level (i.e. the world at large), analogous phenomena are also observed to occur because of the same constraint imposed by the constant quantity of resources (i.e. material and energy). One example of these phenomena is that human beings cannot grow indefinitely. If they do, then how can the world sustain that growth with limited available resources (e.g. foods) to support it?

Another analogous phenomenon is in population growth. Thomas Malthus (1789) predicted that when population growth exceeds agricultural production, the population level is forced to return to subsistence-level conditions. This phenomenon as Malthus (1789) believed is due to the restriction of available resources supporting population growth, which others have claimed to be exploding as the earth becomes over populated.

In a surprising manner, a similar phenomenon can be observed in production processes in which the increase in a single factor of production assuming all other factors constant, results not to more marginal output but to a reduced one. This was attributed by classical economists, Thomas Malthus and David Ricardo, to the decreasing quality of inputs while neoclassical economists attributed this to the disruption in the entire production process as additional units of input are added to a fixed amount of capital. But in view of the Principle of Intrinsic Reality, the phenomenon is viewed intrinsic to the production system in which case infinite amounts of mass or energy cannot be processed with perfect efficiency. Hence, the law of conservation of mass and energy comes into the fore.

It can be deduced from the discussion that there is an underpinning intrinsic reality to the more obvious phenomena in the world from the micro-scale (bacterial growth) to the macro-scale (human growth and production processes).

2.2 Proof Based on Diminishing Marginal Utility

Another way to prove the Principle of Intrinsic Reality is through the diminishing marginal utility observable among humans and perhaps other organisms as well. Marginal utility, as defined in economics, is the gain from an increase or the loss from a decrease in the consumption of goods or services. The diminishing marginal utility is an observable phenomenon in which the utility or satisfaction an individual gains diminishes as he or she consumes more of the same unit of goods or services. Graphically, this can be represented as in Figure 2.

Figure 2: Graphical Representation of Diminishing Marginal Utility

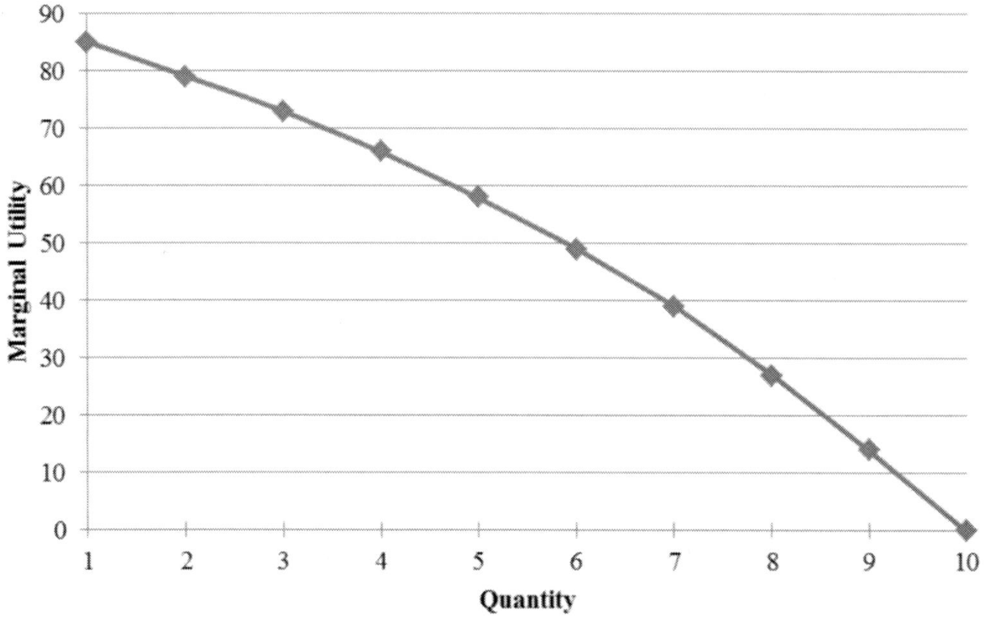

It can be deduced from this phenomenon and in support of the discussion in section 2.1 that human beings cannot grow indefinitely due to the fact that in our intrinsic reality, there is always a limit so as the constancy of available resources (i.e. mass and energy) can support our existence.

This means that in view of our intrinsic reality, we cannot have infinite greed otherwise an upsurge of unintended consequences are bound to happen (i.e. environmental degradation, terrorism and eventually demise of humanity).

Therefore, recognizing our intrinsic reality is the first step towards solving the world's woes and thereby resulting to sustainability and security.

2.3 Expanding the Production Possibility Curve through R&D

The caveat here is that the constant mass or energy (i.e. the limits) leads to trade-offs. Trade-offs is encountered when personal or business decisions are made. In making these decisions, one has to ensure that he or she becomes better off with his or her chosen option. We also realize in our life that we cannot get all that we want and we have to make trade-offs so long as it places us in a better or preferred position. This is more of a philosophical discourse, but it goes to show that the concepts of chemical engineering can be as useful in life and in business as it is in chemical processing.

To tackle these trade-offs, the Production Possibility Frontier in economics can be used. In view of the Principle of Intrinsic Reality where growth is limited by the fixed amount of material and energy, the Production Possibility Frontier needs to shift to the right as shown in Figure 3.

Figure 3. Graphical Representation of Expanding the Production Possibility Curve

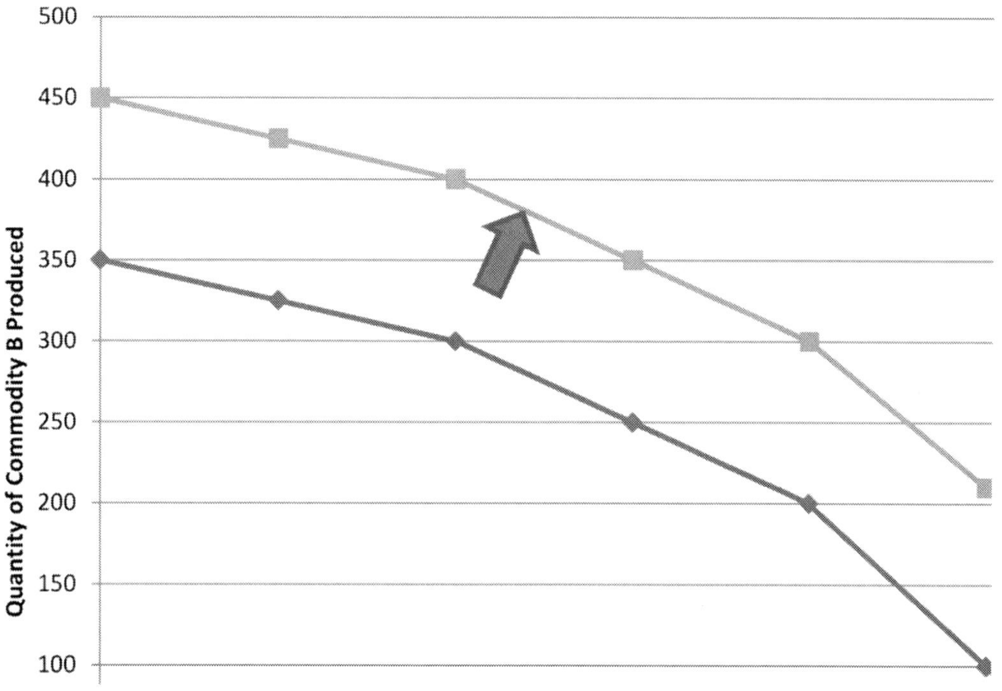

This shift has an implication on the practice of chemical engineering as Research and Development (R&D) is needed to achieve the said shift. R&D enables the economy to produce more of the two commodities (i.e. Commodity A and Commodity B) resulting in economic growth. The key challenge here to Chemical Engineering, particularly on Chemical Engineering R&D, is to look at the bigger picture where economic growth emanating from R&D output is pivoted on sustainability; considering its impact to the environment and the society at large.

3. Paradigm Shift in the Practice of Chemical Engineering

"Money" or "Profit Maximization" has been at the center of the practice of chemical engineering (Smith 2005). Various mathematical optimization methods are used to design systems, processes or components of processes with the end in mind of optimizing profits. This mental model however, will no longer work in our situation today as we face economic, ecological, social and spiritual crisis (Scharmer 2010).

To address this, Scharmer (2010) pointed out seven acupuncture points to shifting capitalism into a regenerative ecosystem economy. Within this purview, the practice of chemical engineering has a greater role to assume. For example, Scharmer (2010) suggested that the design of all production and consumption cycle be completely earth-to-earth, i.e. without necessitating for landfill and that it has to co-evolve with the natural ecosystem. In this regards, the way we do optimization should be revamped into sustainable optimization. By sustainable optimization, it means optimization of the entire system and adding in environmental as well as societal objectives to the equation.

In this definition, two things need to be considered. The first one is systems thinking which emphasizes the understanding of the entire system so the interaction among components will be known and unintended consequences mitigated (Senge 2006). This is a holistic approach to problem solving and finds enormous use in process design.

It should be noted however that the construction of processing plant, whether it is a greenfield or a brownfield project, is rooted on projected market demand. In some instances, economic volatility necessitates companies to experiment with multiple pathways to survive. In this circumstance and within the purview of systems thinking, design thinking comes into the fore. Design thinking is a collective principle emphasizing empathy for users, doing prototyping and tolerance for failure (Kolko 2015). This principle finds enormous use in chemical engineering designs such as in product development, process development, systems improvement and the like. This is also useful for simplifying or humanizing chemical engineering designs with the end in mind of contributing to sustainability. It also has to be noted that of paramount importance in using this tool is imagining the future with the view in mind of status quo and the likely outcome for sustainable development.

The second consideration is the inclusion of environmental and societal objective functions to the objective functions (economic criteria) of profit, cost, energy and yield (Edgar, et al. 2001) in optimization. The environmental objectives can be the impact or risk to environment and the societal objectives can be the impact or risk to society or common good.

These are just few examples of the needed paradigm shift in chemical engineering practice in consideration of the principle of intrinsic reality. But there can be more. A picture of the envisioned new paradigm for the practice of chemical engineering is shown in Figure 4.

Figure 4. New Paradigm in the Practice of Chemical Engineering

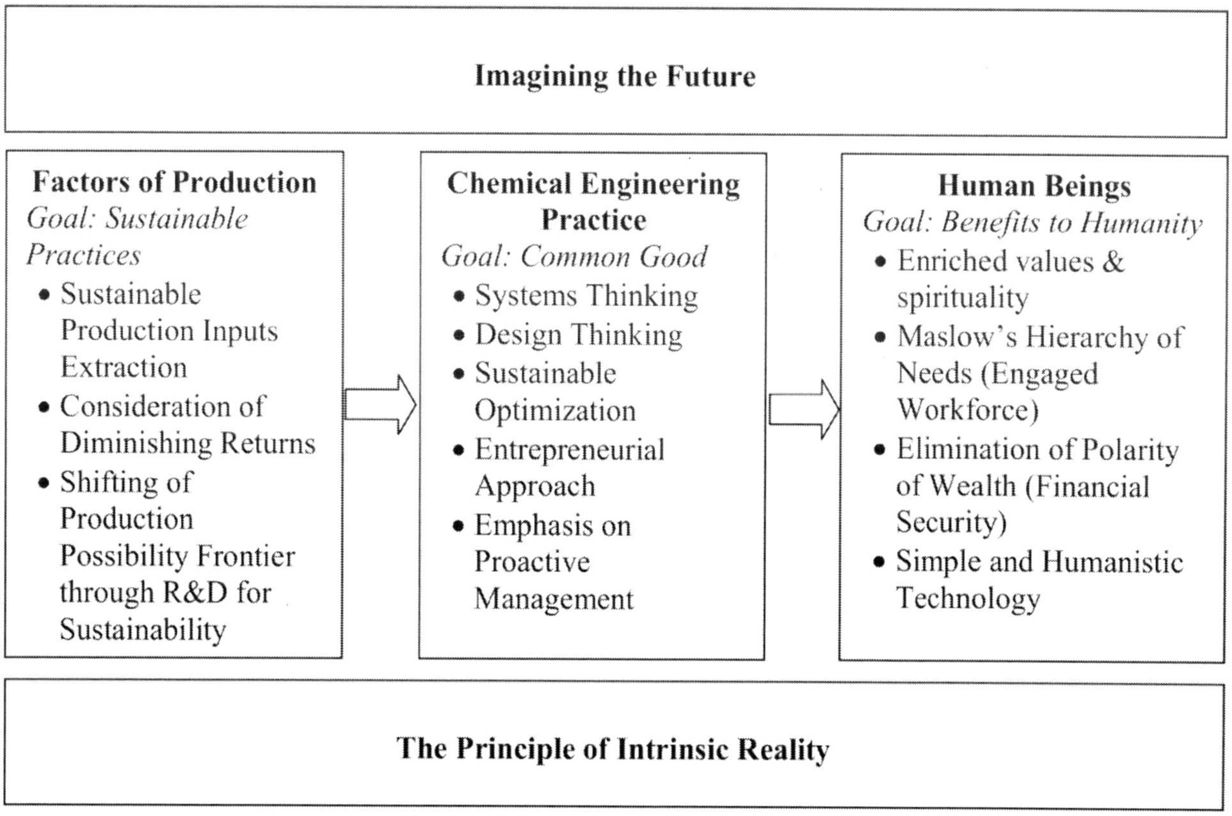

It should be realized that at the core of our economic systems is chemical engineering which can drive the economy into positive direction with the least unintended consequences. Related to that is capitalism in which upgrading of its operating systems can affect the practice of chemical engineering to a substantial degree. A discussion of this is provided in the following section.

4. Wealth Creation and Its Imperative for Sustainability

The development of this paper was inspired by the book "Freakonomics" in which the authors Steven D. Levitt and Steven J. Dubner (2005) asserted that economics can explain almost everything. It came to mind that if economics as a basic science can explain almost everything, then why not chemical engineering which was founded on chemistry, the central science?

Reflecting on it deeper, while explanations clear our doubts, we need more than that so we can tackle the world's woes. As such, this author thought that perhaps combining the two disciplines of chemical engineering and economics can be a potent force towards this end. How this can possibly be done is discussed in this section.

The starting point here is a review of the economic history of the world. In his book "A Farewell to Alms", Gregory Clark (2007) reviewed the economic history of the world in which he divided it into two periods; the period before 1800 and the period after. This is graphically represented in Figure 5.

Figure 5. Shares of Major Economies in World GDP through Time

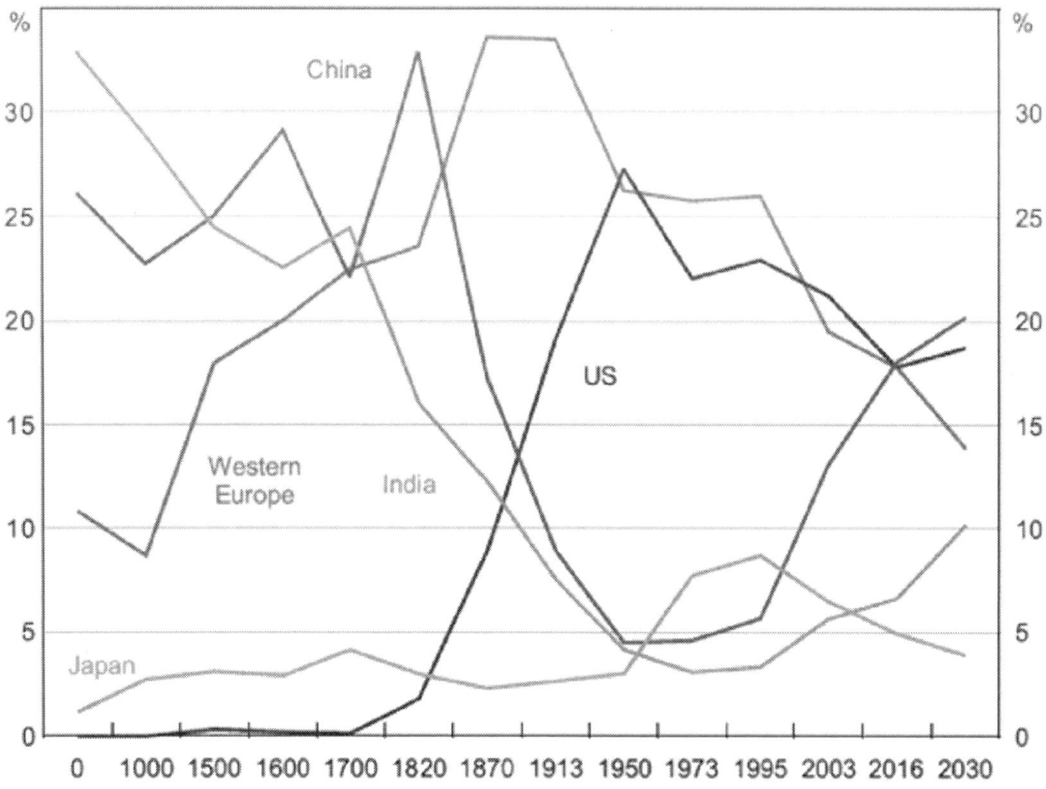

Note: Based on purchasing power parity (PPP) valuation of country GDP

Sources: International Monetary Fund (IMF), *World Economic Outlook* database, April 2011; Maddison (2006)

Thompson (2012) interpreted this graph by stating that "everything to the left of 1800 is an approximation of population distribution around the world and everything to the right of 1800 is a demonstration of productivity divergences around the world." In other words, everything to the left before Industrial Revolution depicts economic power by the sheer size of population and everything to the right after the Industrial Revolution shows the effect of mass production on country's economic output.

Clark (2007) on the other hand called the era before the industrial revolution, i.e. ca. 1800, as Malthusian era in which income per person varied across societies and epochs but growth is not observed. As Clark (2007) explained, this was due to a mechanism called Malthusian trap in which short-term increase in income through advancement in technology is dragged down by population growth. He further explicitly stated:

> "Economics as a discipline arose in the dying decades of the Malthusian era. Classical economics was a brilliantly successful description of this world. But the torrent of goods unleashed by the Industrial Revolution not only created extremes of wealth and poverty across nations, it also undermined the ability of economic theory to explain these differences."

In view of this statement, Figure 5 is reinterpreted in this paper. The contention of this author is that *polarity of wealth* can be gleaned from the graph. *Polarity of wealth* means the uneven concentration of wealth in the global economic system as some parts of the world are richer than others. The polarity can also be observed to shift as some richer parts of the world decline in economic power and some poorer parts of the world rise in economic wealth.

A historical example can be gleaned from Figure 5. India and China up until 1500 CE had 50 and 60 percent share of the world's economy, respectively. In 1 CE, China had a population of almost 60 million, while the area of the present day United States had a population of 680,000. In the late 18th century, the Industrial Revolution rendered countries with large population, just countries with the largest populations.

Looking at North America, it took them 1800 years to overtake China's economic output. However, by 1950, the US economic output was three times as great as China, even though its population is three times less. Comparing China with UK in 1913, the UK had a population of 45 million and China had 437 million but the economic output was almost the same. Analyzing what happened in Europe and later in the United States, Europe was on the lead even before 1800 (Nye 2012).

In 2010 according to data, China overtook Japan as the second biggest economy, though the numbers may not be strictly comparable. Japan's data are seasonally adjusted while those for China were not. Angus Maddison suggested that China and India were the biggest economies in the world for most of the past 2000 years. That is why economists find it a mystery why they fell so far behind and now they are flourishing ('China's economy' 2010). This may not be mysterious considering the *polarity of wealth* as was mentioned earlier and that this polarity can be argued to emanate from the Principle of Intrinsic Reality.

The reasoning here is that the world has constant mass and energy. Assuming that all these mass and energy can be harnessed to create wealth, it could mean that there is a limit for wealth creation. Theoretically, the limit can be expanded indefinitely through R&D but experience demonstrates that this was not achieved yet or getting anywhere near being achieved. As such, the constant mass and energy vis-à-vis the limit for wealth creation can be thought of as the reason for the polarity of wealth, i.e. wealth can only shift from one entity to another.

The implication of this assertion can be demonstrated by the fact that the 1% richest have accumulated more wealth than the rest of the world combined (Oxfam International 2016). The world has never seen 100% of its population to be wealthy. As such, time and again, equitable distribution of wealth is called for and mechanisms to achieve such_should be made a priority in government policies.

In a more concrete way, Scharmer (2010) suggested measures to address this growing inequality which is consistent with the mindset espoused by the Principle of Intrinsic Reality. These measures are as follows:

- Creation of a New Class of Common Property Rights

The creation of common property rights is intended to address the cost to society of private economic undertakings. For example, a private company is extracting natural resources from a certain place. While the private company can profit from that endeavor, the community in that place might suffer the undesirable consequences (e.g. flooding due to forest denudation). To ensure sustainability and institutionalize the voice of future generations in this case, a trust can be established which will own the rights and distribute funds (e.g. from cap-and-trade system for carbon emissions) to citizens (one share per person), communities and the government for investment in public infrastructure.

The trust is an effective means because it does not function like a company (driven by profits) or government (driven by special interest groups) but that it is accountable for long-term sustainability of the specific commons it manage for future generations.

- Creation of a Human Right to Basic Income

One way to ensure equitable distribution of wealth is to create a human right to basic income for every human being. This right when combined with free or affordable access to healthcare and education would create a playing field where everyone has a chance to pursue aspirations and dreams. In other words, it enables individuals to put their real creativity for the benefit of larger community. Another intention of this measure is to free up paid labor (a form of dependence) so that human beings can enhance their creativity and entrepreneurial skills.

- Inclusive Capitalism

To eradicate the *polarity of wealth*, capitalism should be made inclusive so that money and capital flows are directed to all sectors of the economic system. Currently, our money system leads to accumulation of oversupply of money and capital in high financial but low environmental and social gain areas. At the same time, there is an undersupply of money and capital in areas that address societal and community needs. It implies that there is too much money in areas where it is not needed but not enough on areas where it is needed. To address this imbalance, it is recommended to redesign the system so that the current extraction-based financial system is replaced by institutions which can support the real economy and thereby serve the ecology and community.

It should be noted that shifting of wealth is also observed in various geographical regions. When this occurs, geo-political instability can result. This goes to show that if we want lasting stability and security within the limits of our intrinsic reality, then we should not tolerate unabated greed. Otherwise, *polarity of wealth* will continue to persist and we cannot realize our full potential as individuals and collaborators for common good.

5. Way Forward...

Recognizing our intrinsic reality leads us to greater understanding of our very nature bounded by the constancy of mass and energy. It shed light on why growth stabilizes after addition of more inputs. Within this purview, trade-offs are also encountered which necessitates for R&D. R&D initiatives however should be underpinned on sustainability.

Looking at the practice of chemical engineering from the lens of intrinsic reality, the approach should be futuristic while considering relevant aspects related to factors of production and human beings.

The bigger picture shows that Engineers particularly Chemical Engineers are major players in driving the real economy. As Scharmer (2010) aptly puts it, "The essence of technology and knowledge creation, and the ultimate driving force of the economy, is the same: it is the human capacity to create." As such, if sustainability is to be achieved, Chemical Engineers have a greater responsibility to innovate for regenerative and social technologies. In this way, we can stem the polarity of wealth by providing affordable technology to enable disadvantaged individuals enhance their creativity, innovation and entrepreneurship. This can help them partake in the benefits of inclusive capitalism.

As demonstrated in this paper, the synergy emanating from the convergence of Chemical Engineering and Economics can powerfully tackle the world's woes. As such, there is a good reason for chemical engineers and economists alike to collaborate particularly on sustainability initiatives.

6. References

[1] 'China's economy overtakes Japan's in real terms' 2010, *The Economist*, 16 August, viewed 5 March 2016, http://www.economist.com/node/16834943

[2] Clark, G 2007, *A Farewell to Alms: A Brief Economic History of the World,* Princeton University Press, Princeton, NJ.

[3] Edgar, T, Himmelblau D & Lasdon, L 2001, *Optimization of Chemical Processes*, McGraw-Hill, New York.

[4] Huang, Y & Wang, B 2011, 'From the Asian Miracle to an Asian Century? Economic Transformation in the 2000s and Prospects for the 2010s', in H Gerard & J Kearns (eds), *Proceedings of the Reserve Bank of Australia annual conference*, Reserve Bank of Australia, Sydney, viewed 30 January 2016,

http://www.rba.gov.au/publications/confs/2011/huang-wang.html

[5] Kolko, J 2015, 'Design Thinking Comes of Age', *Harvard Business Review*, pp. 66-71, viewed 24 February 2016, https://hbr.org/2015/09/design-thinking-comes-of-age

[6] K.N.C. 2012, 'Mis-charting economic history: more 2,000 years in a single graphic', *The Economist blog*, web log post, 20 June, viewed 29 February 2016,

http://www.economist.com/blogs/graphicdetail/2012/06/mis-charting-economic-history

[7] Levitt, S & Dubner, S 2005, *Freakonomics: A Rogue Economist Explores the Hidden Side of Everything*, William Morrow, New York, NY.

[8] Malthus, T 1798, *An Essay on the Principle of Population*, J. Johnson in St. Paul's Church-yard, London.

[9] Maslow, A 1943, 'A Theory of Human Motivation', *Psychological Review*, vol. 50, pp. 370-396, viewed 24 February 2016, http://psychclassics.yorku.ca/Maslow/motivation.htm

[10] Monod, J 1949, 'The Growth of Bacterial Cultures', *Annual Review of Microbiology*, vol. 3, pp. 371-394, viewed 2 February 2016,

http://www.annualreviews.org/doi/abs/10.1146/annurev.mi.03.100149.002103

[11] Nye, J 2012, 'Fascinating new graph shows the 'economic history of the world since Jesus'', *MailOnline*, 23 June, viewed 15 November 2015,

http://www.dailymail.co.uk/news/article-2163610/Fascinating-new-graph-shows-economic-history-world-Jesus.html

[12] Oxfam International 2016, *An economy for the 1%, how privilege and power in the economy drive extreme inequality and how this can be stopped*, Oxford, viewed 5 March 2016,

https://www.oxfam.org/sites/www.oxfam.org/files/file_attachments/bp210-economy-one-percent-tax-havens-180116-en_0.pdf

[13] Samuelson, P & Nordhaus W 2009, *Economics*, McGraw-Hill Higher Education, New York, NY.

[14] Scharmer, O 2010, 'Seven Acupuncture Points for Shifting Capitalism to Create a Regenerative Ecosystem Economy', *Oxford Leadership Journal*, vol. 1, pp. 1-21, viewed 24 February 2016, http://www.oxfordleadership.com/journal/vol1_issue3/scharmer.pdf

[15] Senge, P 2006, *The Fifth Discipline: The Art & Practice of the Learning Organization*, Doubleday/Currency, New York.

[16] Smith, R 2005, *Chemical Process Design and Integration*, John Wiley & Sons Ltd., West Sussex, England.

[17] Thompson, D 2012, 'The economic history of the last 2,000 years in 1 little graph', *The Atlantic blog*, web log post, 19 June, viewed 29 February 2016, http://www.theatlantic.com/business/archive/2012/06/the-economic-history-of-the-last-2-000-years-in-1-little-graph/258676/

Acknowledgment

The author would like to thank Hyundai Engineering and Construction Co. Ltd. for the support in this endeavor. The author is also thankful to Ms. Sofia Ayu for meticulously editing this manuscript and Mr. Marcial N. Barroga for the poster design.

442106 Effects of Mono-Ethylene Glycol on Three-Phase Low Liquid Loading Flow Characteristics in Near-Horizontal Pipelines

Abstract:

Effects of Mono-Ethylene Glycol on Three-Phase Low Liquid Loading Flow Characteristics in Near-Horizontal Pipelines

Mono-ethylene glycol (MEG) is commonly used in deep water gas production systems as a hydrate inhibitor. However, MEG mixing in multiphase flow and its effects on flow parameters are not well understood. This paper provides with comprehensive data for three-phase stratified flow in a 6-*in* ID pipe, with MEG in the aqueous phase. In addition, the prediction performances of the commonly used predictive tools in the industry are provided.

The experimental study is conducted using a 6-*in* ID facility to investigate the effects of the presence of MEG on three-phase stratified wavy flow in horizontal pipelines under low liquid loading conditions, commonly observed in wet gas pipelines. The analyzed flow characteristics include wave pattern, liquid holdup, aqueous phase fraction, pressure gradient, and wetted wall fraction.

The experimental range covers superficial gas velocity, v_{Sg}, values of 8-23 *m/s*, superficial liquid velocity, v_{SL}, values of 1-2 *cm/s*, and inlet liquid stream aqueous phase fraction, WC_{MEG}, values of 0-100%. Experiments are conducted with 50 *wt*% of MEG in the aqueous phase, and the results are compared to the case with no MEG, presented by Karami *et al.* (2015). Differential pressure transmitters, a quick closing valve and pigging system, and a high speed camera are used for data acquisition. The trends in the resulting data with respect to input parameters are investigated. The performances of commonly used models and predictive tools are compared to liquid holdup, pressure gradient and aqueous phase fraction experimental results.

The observed wave patterns include stratified smooth and stratified wavy with 2-D waves, 3-D waves, roll waves, and atomization flow, with transition boundaries varying with the liquid phase composition. The trends of pressure gradient, liquid holdup, and aqueous phase fraction with respect to v_{Sg}, v_{SL}, WC_{MEG}, and MEG *wt%* are observed, and physical justifications are provided. The predictions of a commercial multiphase flow simulator (OLGA 7.1), an in-house software (TUFFP unified model v. 2012), Beggs and Brill (1973) correlation, and Taitel and Dukler (1976) and Xiao *et al.* (1990) models are compared with the acquired experimental data. The liquid holdup experimental data are under-predicted by all the models, especially for higher WC_{MEG} values. However, the results from OLGA 7.1 and Xiao *et al.* (1990) model are in better agreement with experimental data. The three-phase aqueous phase fraction trends are not predicted well. The complex nature of liquid-liquid interactions in three-phase low liquid loading flow causes higher uncertainties in predictions.

References:

Beggs, D. H., & Brill, J. P. (1973). A study of two-phase flow in inclined pipes. *Journal of Petroleum technology*, **25**(05), 607-617.

Karami, H., Torres, C., Pereyra, & E., Sarica, C.: "Experimental investigation of Three-Phase Low Liquid Loading Flow", SPE-174926-MS, presented at *2015 SPE ATCE*, Houston, TX, 28-30 September, 2015.

Taitel, Y., & Dukler, A. E. (1976). A model for predicting flow regime transitions in horizontal and near horizontal gas liquid flow. *AIChE Journal*, **22**(1), 47-55.

Xiao, J. J., Shoham, O., & Brill, J. P. (1990, January). A comprehensive mechanistic model for two-phase flow in pipelines. In *SPE Annual Technical Conference and Exhibition*. Society of Petroleum Engineers.

Copyright © American Institute of Chemical Engineers. All rights reserved.

443637 Transport of Water and Oil Droplets Entrained in Gas Phase for Larger Diameter Air-Oil-Water Low Liquid Loading Flow in Horizontal and Near-Horizontal Pipe

Abstract:

This paper describes a study on entrainment of oil and water in gas phase in case of air-oil-water three phase low liquid loading flows. The experiments were carried out in a 6-in diameter pipe at inclinations of -2, 0 and +2 ° from the horizontal. An isokinetic sampling probe system was used to measure flux of entrained droplets. The distribution of entrained drops is highly asymmetric in horizontal flow. Hence, droplet flux was measured at different locations along the vertical axis of the pipe. The liquid superficial velocity (v_{SL}) and gas superficial velocity (v_{SG}) were varied between 0.005 m/s to 2.5 m/s and 10 m/s to 23 m/s, respectively. All the three phase experiments were conducted for oil continuous phase and water cut of 0, 10, 20, 40 and 100 %.

The entrainment fraction was highly sensitive to gas flow rate. The effect of liquid flow rate and inclination was less significant. Although entrainment fraction tended to increase with change in inclination from -2 to +2°, the effect diminished as gas flow rate increased. Entrainment of water was higher in air-oil-water three phase flow compared to air-water two-phase flow. Thus, entrainment of water which is the dispersed phase was enhanced by the oil continuous phase. The fraction of water in the entrained liquid decreased with increasing distance from bottom of the pipe due to higher settling velocity of water compared to that of oil.

An approach to modeling of the water and oil entrainment is presented assuming homogenous oil-water mixing in the liquid phase. This approach assumes stratified atomization flow pattern, which is predominant flow pattern for low liquid loading flow conditions. The approach suggested in current study for determination of entrainment fraction in low-liquid loading fairs better than the available correlations in describing the functional dependence of entrainment fraction on superficial liquid velocity. This approach is extended to three-phase flow by assuming that the deposition of the entrained water and oil drops takes place independent of each other. Uniform distribution of water in oil continuous phase is assumed to predict rate of atomization, which matches the experimental observations except at lower gas velocity. For low gas flow rate (v_{SG} < 19 m/s) investigated in this study, the proposed correlation over predicts amount of water entrained in the gas phase.

Copyright © American Institute of Chemical Engineers. All rights reserved.

TEMPERATURE INFLUENCE ON GAS-LIQUID FLOW REGIME TRANSITIONS

Luisa Lezama, Valeriy Koukhtiev, Alexey Pushkin and Brent C. Houchens
Landmark-Halliburton, Houston, Texas, USA.

Abstract

The impact of temperature variations on the onset of multiphase flow regime transitions is explored over parameter spaces typical of production and pipeline transport. Influences on flow regime maps resulting from changes in liquid viscosity and gas density, the two most temperature-sensitive parameters, are investigated. For significant increases in the liquid viscosity, the most dramatic impacts on the vertical-upward flow map are the increased dominance of the dispersed bubble and churn regimes. As the gas density increases and approaches that of the liquid density, the annular flow regime becomes more dominant for both vertical-upward and horizontal flows. A case study by means of a one-dimensional (1D) steady state nodal analysis was performed on a 3300 m, vertical, heavy oil, onshore well to demonstrate the impact of various thermal model assumptions. The bottom-hole flowing pressure was predicted and compared to field data, with matched wellhead pressure and production rates. Surprisingly, this common procedure of matching flow rates can significantly mask errors. While the bottom-hole flowing pressures varied by less than 4% with three different thermal model assumptions, it was possible to observe downhole locations with nearly 13% variations in average density. The depths of regime transitions were also shifted by as much as 150 m.

Introduction

Mechanistic models that identify flow regimes and predict pressure drop in two-phase, gas-liquid flows are widely used in flow assurance calculations during production and pipeline transport. Regime transitions are influenced by fluid properties, which are functions of temperature, pressure, and composition. In some production and transport situations, the oil and gas experience significant temperature variations. These variations can cause flow regime transitions and thus impact pressure-drop and liquid-holdup predictions. Therefore, the effect of temperature on regime transitions should be considered to accurately design and maintain systems for multiphase flow transport and separation.

Regime transitions are particularly impacted in deepwater wells, where much of the vertical lift portion is submerged in seawater at temperatures as cold as 4°C. Regime transitions in seafloor pipelines, jumpers, and vertical-lift production tubing are modified by the significant heat transfer to the low-temperature seawater. It is not uncommon for flow conditions to place flow very near a regime transition. Such a situation was recently shown for a deepwater riser flow that experienced dispersed bubble flow very near the transition to slug flow [1]. The shift in the dispersed bubble to slug transition as liquid viscosity increases is discussed.

In vertical-lift sections, the depth at which gas is released from liquids at the bubble point can be moved significantly upstream or downstream in the wellbore. This moves the onset of regime transitions from single-phase liquid to bubbly and dispersed bubble flow for oil-producing wells containing dissolved gas. This is true even for onshore wells, as is demonstrated by the case study of a heavy oil well presented here. The subsequent transition to slug flow is offset proportionally because of the shift in onset of bubbly or dispersed bubble flow.

Similarly, pipelines at extreme latitudes experience very cold winters. In addition to impacting the equilibrium flow regime, the frictional pressure drop is significantly increased by cooling the production fluids, which can dramatically increase the liquid viscosity or even transition the fluids below their pour point. Waxy crude oils are particularly impacted [2]. In near-horizontal pipelines, the transition from stratified smooth to stratified wavy flow is modified through the influence of temperature on viscosity. This is significant for corrosion issues, where stratified wavy flow is preferred over stratified smooth flow. Simultaneously, the impact on viscosity has a significant influence on the pressure drop.

Models and Methods

Two-phase, gas-liquid pipe flow regimes are first determined using regime transition functions similar to those presented by Barnea [3]. The flow regime model can identify a quiescent mixture, single-phase gas or single-phase liquid flows, or multiphase flow regimes. The multiphase regimes include dispersed bubble, bubbly, slug, elongated bubble, churn, annular, stratified smooth, and stratified wavy flow. Once the regime is identified, 1D momentum equations are solved with closure relations for the liquid holdup and pressure drop similar to the method of Gomez *et al.* [4]. For the purpose of brevity, the reader is referred to these two references for details of mechanistic modeling.

Dependencies of the Mechanistic Regime Transition Functions

The regime transition functions for two-phase, gas-liquid pipe flow depend on ten input parameters, including the superficial gas and liquid velocities v_{SG} and v_{SL}, the gas and liquid densities ρ_G and ρ_L, the gas and liquid viscosities μ_G and μ_L, the surface tension of the liquid in contact with the gas σ_L, the pipe diameter D, the pipe inclination angle θ measured from horizontal, and the pipe wall roughness ε_r. Regime identification depends strongly on the superficial gas and liquid velocities and the pipe inclination angle, as demonstrated by the experiments of Shoham [5] on air and water. Though limited experiments exist for other fluid combinations, a review of the regime transition functions suggests that the pipe diameter, liquid viscosity (by means of the friction factor), and density difference play secondary roles in determining the equilibrium regime. Among these secondary dependencies, the liquid viscosity μ_L and density difference $\rho_L - \rho_G$ can both vary significantly with temperature.

Temperature-Sensitivity Study Isolated to Liquid Viscosity and then Gas Density

The effect of temperature on the regime transition functions is studied through varying first the liquid viscosity and then the gas density for both vertical-upward and horizontal pipe flows. The regime transition functions of Barnea, generalized to account for cases in which the liquid and gas densities are similar, were coded and sweeps over typical ranges of superficial gas and liquid velocities were performed to produce flow regime maps. The influences were isolated to investigate the impact.

Case Study: Heavy Oil and Gas Well

Regime transition functions were implemented in NETool™ simulator as part of the LMK-1 multiphase flow model. A case study was conducted for a vertical well producing heavy oil and gas. This well was modeled based on matching to the field data presented by Chierici *et al.* [6]. Three different wellbore temperature scenarios were considered.

Results: Influence of Liquid Viscosity and Gas Density on Regime Transitions

First, the influence of varying only the liquid viscosity on the regime transition functions is demonstrated. Next, the influence of varying only the gas density is investigated. These represent isolation of the two most highly temperature-dependent input parameters. The other parameters are set to representative values for situations that could exist along wellbores producing oil and natural gas.

Influence of Liquid Viscosity on Regime Transition Functions

The liquid viscosity μ_L can vary significantly in a wellbore as a result of temperature variations from the bottom hole to the wellhead. Across a wide range of oil API values, the variation is even more significant. For example, from light oil at warm temperatures to cold heavy oil, it is possible for the liquid viscosity to vary from 0.02 cP (2×10^{-5} Pa-s) up to 5000 cP (5.0 Pa-s), respectively.

Regime maps for vertical and horizontal flow are shown in Figure 1 in the v_{SG} - v_{SL} space over a range of liquid viscosities from 1×10^{-4} to 1.0 Pa-s. All other input parameters are held constant at values typical for oil and natural gas of $\mu_G = 1\times10^{-5}$ Pa-s, $\rho_L = 700$ kg/m^3, $\rho_G = 200$ kg/m^3, and $\sigma_L = 0.03$ N/m. A representative diameter of 0.1 m was used, and the pipe was assumed smooth. Note that holding the densities and gas viscosity fixed is somewhat artificial but allows only the effect of liquid viscosity to be demonstrated.

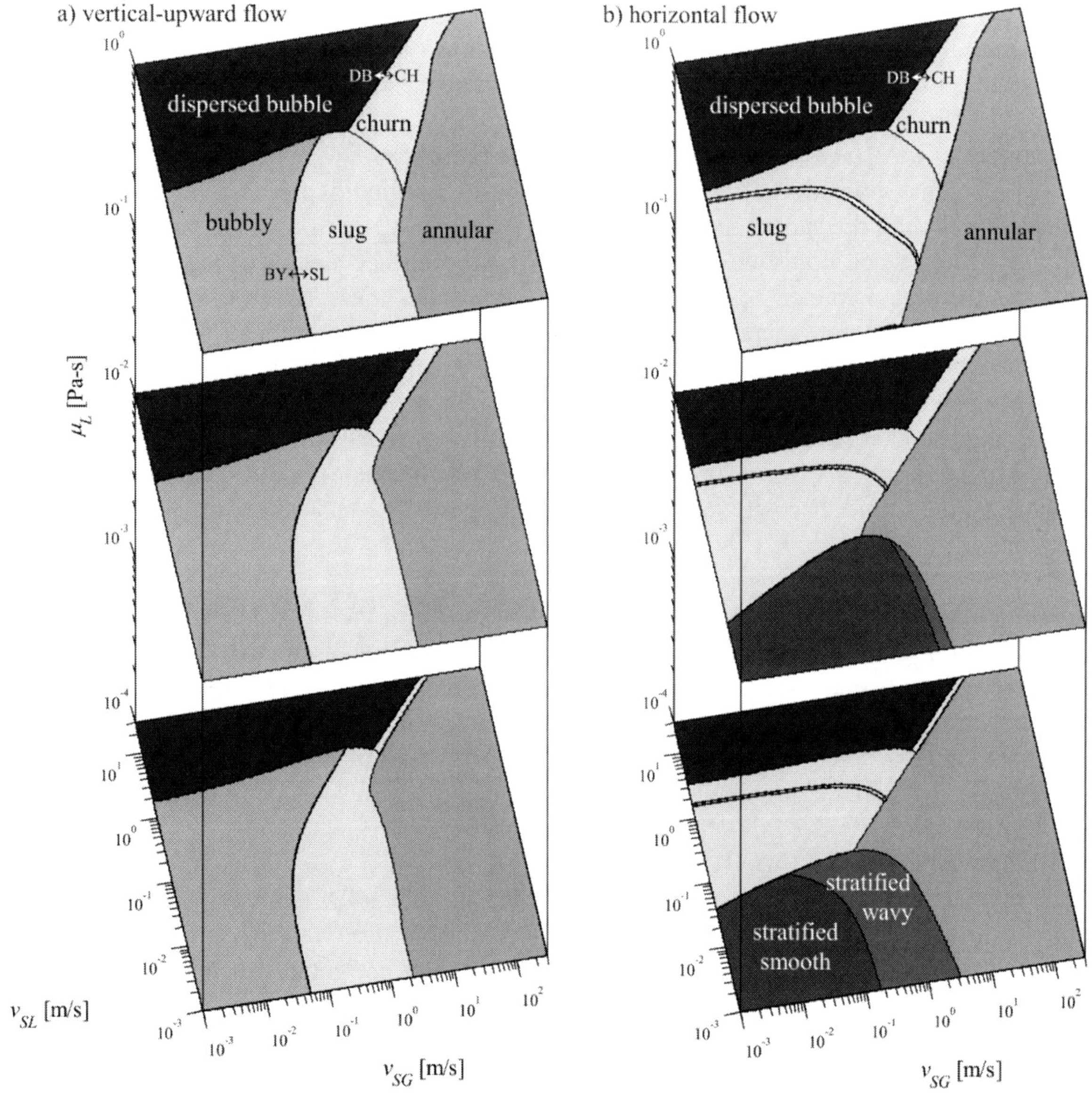

Figure 1. Influence of liquid viscosity on the regime maps for gas-oil systems with a) vertical upward flow and b) horizontal flow. All other parameters are held constant with values $D = 0.1$ m, $\mu_G = 1.0\times10^{-5}$ Pa-s, $\rho_L = 700$ kg/m^3, $\rho_G = 200$ kg/m^3, and $\sigma_L = 0.03$ N/m for a smooth pipe. The thin yellow bands within the slug flow region represent the elongated bubble regime.

It is first worth noting that two of the boundaries do not move. In both vertical-upward and horizontal flows, the dispersed bubble to churn transition (DB↔CH) does not vary with friction factor and thus does not depend on μ_L. The DB↔CH transition occurs when the no-slip gas-void fraction $\alpha_{G_{NS}} = v_{SG} / v_M$ exceeds 0.52, the maximum packing factor for simple-cubic spheres [7]. Similarly, for vertical-upward flow, the bubbly to slug (BY↔SL) transition occurs when the gas-void fraction exceeds 0.25. This condition considers slip between the phases but does not vary with μ_L.

As the liquid viscosity increases, the dispersed bubble regime becomes stable at lower superficial liquid velocities for both vertical and horizontal flows. This is a result of the decrease in the maximum allowable dispersed bubble diameter $d_{DB,\max}$, which represents the break-up resistant diameter for a specified level of turbulence. This varies as $d_{DB,\max} \propto f_{DB}^{-2/5}$, where f_{DB} is the dispersed bubble friction factor [8]. This friction factor depends on the effective dispersed bubble viscosity μ_{DB}. There are several model choices for μ_{DB}, but the most consistent has been shown to be a volume-weighted average of the gas and liquid viscosities [9]. There are also several models for f_{DB}, but independent of the choice, f_{DB} increases with increasing μ_{DB}. Thus, as μ_{DB} increases in Figure 1, f_{DB} increases and $d_{DB,\max}$ decreases if all other variables are held constant. The value of $d_{DB,\max}$ is compared to two diameters. The first, d_{CD}, sets the deformation-resistant limit to maintain spherical dispersed bubbles, which is required by definition of the regime. The subscript CD indicates "critical deformation." The second, d_{CB}, is relevant for horizontal and inclined pipes and determines when the buoyant forces will overwhelm the turbulent mixing and result in creaming of the bubbles at the top wall of the pipe. The subscript CB indicates "critical buoyant." These two bubble diameters have been previously described in detail [9].

In vertical flow, the regime transition function that compares $d_{DB,\max}$ to d_{CD} is of interest in the DB↔BY and DB↔SL transitions. The regime transition function $d_{DB,\max} < d_{CD}$ is a necessary but not sufficient condition for the existence of stable dispersed bubble flow. This can alternatively be expressed as $d_{CD} - d_{DB,\max} > 0$. Because d_{CD} depends only on ρ_G, ρ_L, and σ_L, it is a constant value for all of the cases encompassed by Figure 1. As $d_{DB,\max}$ decreases with increasing μ_L (and corresponding increasing μ_{DB}), the condition $d_{CD} - d_{DB,\max} > 0$ is satisfied over more of the regime map and dispersed bubble flow becomes more dominant.

In horizontal flow, both the deformation and buoyant creaming regime transition functions play a role in the DB↔SL transition. When comparing $d_{DB,\max}$ to d_{CB}, the friction factor alone does not explain the transition, as d_{CB} varies with both f_{DB} and the average mixture velocity, v_M, which can be calculated from $v_M = v_{SG} + v_{SL}$. It should be noted that $d_{DB,\max}$ also depends on the superficial velocities and v_M, so it is not generally sufficient to consider only the role of f_{DB} on $d_{DB,\max}$. Nevertheless, in the high-liquid-viscosity cases shown in the horizontal flow of Figure 1, it is the critical bubble deformation size ($d_{CD} - d_{DB,\max} > 0$) regime transition function that controls the evolution of the dispersed bubble regime boundary with slug flow.

Figure 1 shows that the churn to slug (CH↔SL) and churn to annular (CH↔AN) transitions shift in favor of churn flow as μ_L increases for both inclination angles. Multiple regime transition functions are involved. First, the annular to intermittent transitions must be considered, where intermittent flow includes slug, elongated bubble, and churn. The annular to intermittent transition occurs when either the annular film becomes unstable because of backflow of the liquid or when there is so much liquid in the film that it blocks the gas core. The regime transition is based on the Lockhart and Martinelli parameter X for separated flows, which represents the ratio of the frictional pressure gradient for liquid flowing alone in the pipe to that for gas flowing alone. The value of X increases with increasing liquid viscosity. With all other factors constant, the liquid holdup increases with X, making blockage of the gas core more likely.

Once the flow is determined to be intermittent, the slug to churn transition is considered. This occurs when the gas holdup resulting from bubbles *within the liquid slug* reach a critically high level, typically taken as 0.52 to match the packing factor for simple-cubic spheres. At this point, the bubbles in the liquid slug begin to coalesce, and the continuous structure of the liquid slug is destroyed. The gas holdup within the liquid slug is proposed by Brauner and Barnea to follow the same mechanisms of transition as that of gas bubbles in dispersed bubble flow [10]. The transition depends on the mixture friction factor f_M in the slug, defined in the same manner as f_{DB}, and the average mixture velocity v_M. The exact dependence is determined by the smaller of d_{CD} and d_{CB}, but in either case, as the viscosity and friction factor increase, the transition to churn will occur at lower mixture velocities and thus the churn regime consumes more territory from the slug regime as the liquid viscosity is increased.

For the relatively high gas density represented in Figure 1, it is also worth noting that the churn regime is much smaller than for cases such as air and water at 20°C and 1 atmosphere. For a lower gas density, the churn regime typically completely separates the slug and annular flow regions for vertical-upward flow (see, for example, Figure 2 for $\rho_G = 5\ \text{kg/m}^3$).

The most interesting trend is that for stratified flow in a horizontal pipe. As the liquid viscosity increases, the stratified regimes shrink and are replaced primarily by slug flow, and to a lesser extent, annular flow. This is a much studied effect [11, 12] and occurs because the equilibrium liquid holdup increases dramatically with viscosity [13]. The stratified to non-stratified regime transition used here is that of Taitel and Dukler [14] and is based on inviscid Kelvin-Helmholtz theory. Thus, it might be expected that the reduced stability of stratified flow with increasing μ_L is a result of the incorrect application of inviscid theory when the liquid viscosity is high. However, Barnea and Taitel showed that, somewhat paradoxically, the inviscid and viscous predictions are very similar for high-viscosity fluids and instead diverge significantly for low-viscosity cases [12]. In their original work, Taitel and Dukler [14] made several assumptions, later described by Barnea [13]. To simplify their analysis, they assumed $\rho_G H_L \ll \rho_L \alpha_G$ and $v_L \ll v_G$. They also included a factor of $1 - h_L / D$ to account for finite wave height. Removing this factor has been shown to greatly improve the accuracy of the predicted liquid height h_L for high-viscosity liquids. The full evaluation is beyond the scope of this paper, but a review of predictions by other groups shows that the stratified to non-stratified regime boundary indeed occurs at decreasing values of v_{SL} for increasing μ_L (see, for example, Figure 3 in Lin and Hanratty for channel flow [11] and Figure 1 in Barnea and Taitel [12]). However, the shift in this regime boundary with increasing μ_L, as observed in Figure 1, is clearly too sensitive because of *i)* the use of the Taitel and Dukler criteria, which includes the factor $1 - h_L / D$ that has been shown to be inaccurate and *ii)* the simplifying assumptions above, both of which are violated for some of the space covered in Figure 1.

Influence of Gas Density on Regime Transition Functions

The gas density ρ_G also varies significantly in a wellbore because of temperature and pressure variations from the bottom hole to the wellhead. Gas densities from approximately 0.1 to 600 kg/m^3 are possible if a range of wells are considered. Regime maps for vertical and horizontal flow are shown in Figure 2 in the v_{SG} - v_{SL} space over a range of gas densities from 5.0 to 500.0 kg/m^3. All other input parameters are held constant with $D = 0.1$ m, $\mu_L = 1\times10^{-3}$, $\mu_G = 1\times10^{-5}$ Pa-s, $\rho_L = 700\ \text{kg/m}^3$, and $\sigma_L = 0.03$ N/m for a smooth pipe. Note that the cases with $\rho_G = 200\ \text{kg/m}^3$ could be added to Figure 1 at a level of $\mu_L = 1\times10^{-3}$ Pa-s.

As in Figure 1, the dispersed bubble to churn transition is not impacted at either inclination angle. The bubbly to slug transition for vertical-upward flow changes only marginally as a result of slip over the range of gas densities.

For both inclination angles, the primary impact is on the annular to intermittent transition. As the gas density increases, the annular regime becomes increasingly stable at lower superficial gas velocities, consuming domain from the churn, slug, and stratified (for horizontal pipes) regimes.

For vertical-upward flow, the impact on the regime transition function is complex, as both X and Y are influenced, where Y is the relative magnitude of gravity on the liquid with respect to the frictional pressure drop that would exist if gas flowed alone in the pipe. With all other factors constant, increasing ρ_G decreases both X and Y, but the net effect is the tendency to favor (stabilize) annular flow over the intermittent regimes of slug and churn. This is a logical result, as a gas core with a density approaching that of the liquid film would be expected to be stable in an annular configuration.

For horizontal flow, the mechanism is more straightforward, as $Y = 0$ because gravity acts perpendicular to the flow direction. In this case, increasing ρ_G still decreases X, leading to lower equilibrium liquid holdup and a more stable annular flow. Again, a gas density approaching that of liquid explains the ability of annular flow to displace the stratified flow territory.

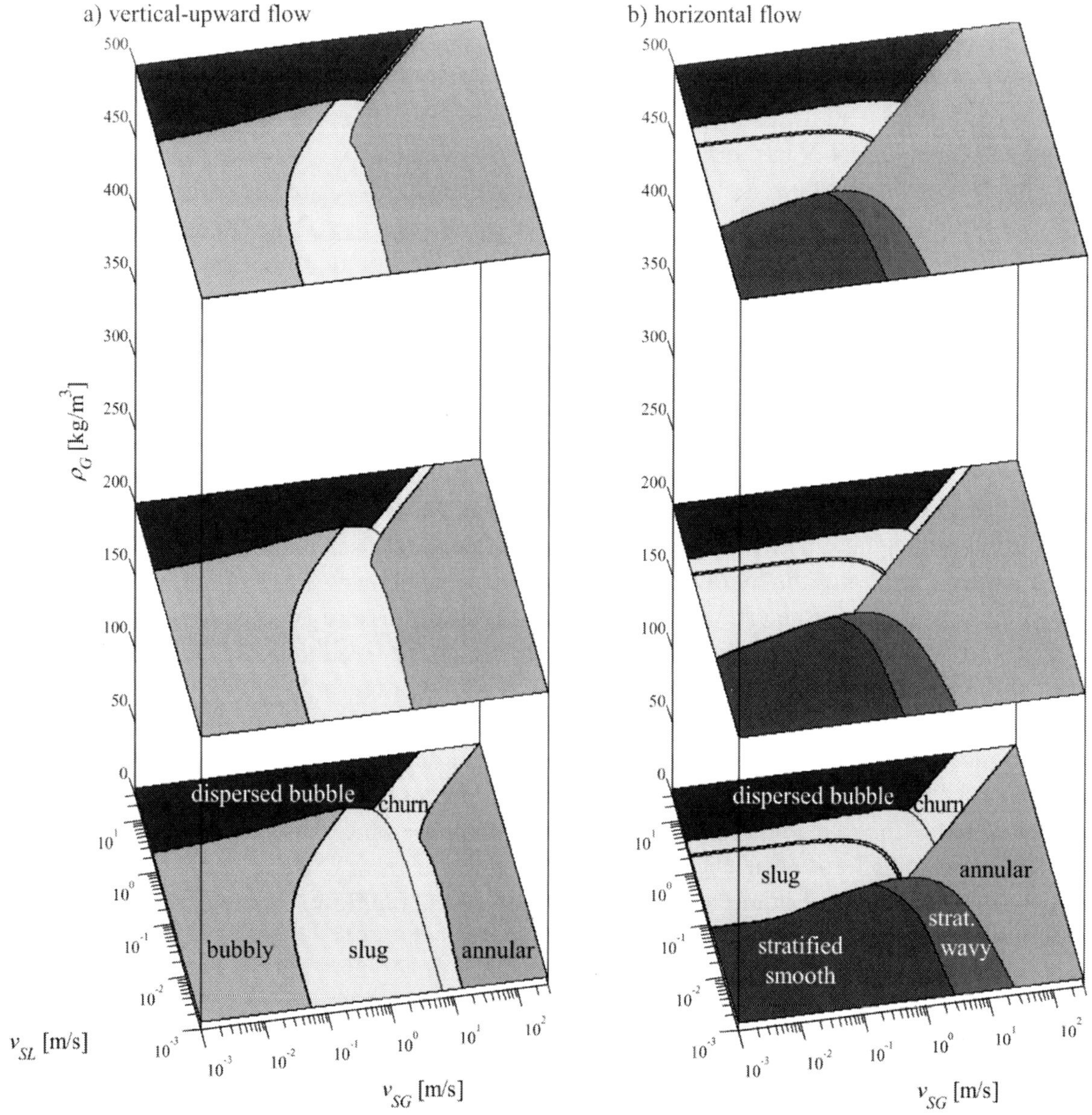

Figure 2. Influence of gas density on the regime maps for gas-oil systems with a) vertical upward flow and b) horizontal flow. All other parameters are held constant with values $D = 0.1$ m, $\mu_L = 1.0\times10^{-3}$ Pa-s, $\mu_G = 1.0\times10^{-5}$ Pa-s, $\rho_L = 700$ kg/m^3, and $\sigma_L = 0.03$ N/m for a smooth pipe. The thin yellow bands within the slug flow region represent the elongated bubble regime.

Results: Case Study of a Vertical Well Producing Heavy Oil and Gas

The previously described influences of temperature on liquid viscosity and gas density are expected to play important roles in wellbore modeling, particularly when the temperature and composition change significantly along the wellbore. To investigate, a case study is presented for the production of heavy oil in a well in which the bubble point pressure is reached within the wellbore.

Well Characteristics

Chierici *et al.* present field data from 31 oil- and gas-producing wells [6]. Well Case No. 2 was selected for this study because *i)* the oil from that well had a very high API, *ii)* the wellbore experienced a significant temperature variation from bottom hole to wellhead, and *iii)* two regime transitions were observed. This 3300 m vertical well produced heavy crude oil with an API gravity of 8.3 at an oil flow rate of 930.9 STB/day with a gas-oil-ratio (GOR) of 145.4 SCF/STB and a very small water cut of 0.15%. The fluids were produced through 5-inch diameter tubing (0.127 m). During production, the measured wellhead and bottom-hole temperatures were 134.6°F (57°C) and 228.2°F (109°C), respectively. The measured wellhead and bottom-hole flowing pressures were 209 psi (1.441 MPa) and 4,763 psi (32.84 MPa), respectively.

This field data set is particularly useful because characteristics of the reservoir oil at the bubble point are provided. These are reported as the bubble point pressure $P_B = 796.5$ psi (5491.7 kPa), the dissolved gas-to-oil ratio (solution GOR) $R_S = 145.4$ SCF/STB measured in saturated reservoir oil compared to stock tank oil (STB), the formation volume factor for saturated oil $B_0 = 1.153$ RB/STB, and the viscosity for saturated oil $\mu_0 = 77.2$ cP. Additionally, the specific gravity of gas relative to air is given as 1.705, and the total dissolved solids make up 63 g/L of produced fluids. There is no method to account for the dissolved solids in the model which, if uniformly distributed along the wellbore at the reported density, could produce a maximum of a 6.5% change in the hydrostatic pressure at the bottom hole, assuming the solid density was much greater than the liquid density. If the solids were the same density as the liquid, then they would account for 6.5% of the volume in the wellbore. The most likely reality is somewhere between these extremes, with the solids accounting for a few percent of the volume and also modifying the hydrostatic pressure by a few percent. It will be shown that, if the flow rates are matched and only the error in bottom-hole flowing pressure is considered, many of the more significant errors caused by temperature variation along the wellbore are masked by competing influences. Thus, the absolute correctness of the overall bottom-hole flowing pressure is of little interest here. Therefore, the influence of solids will be neglected throughout this case study.

The measured depths—identical to the true vertical depths for a vertical well—at which regime transitions occur are also given in the field data. The depths are given first for the transition from single-phase liquid to what the authors call either "bubble or plug" flow, and then for the transition from "bubble or plug" to what the authors call "slug or froth." Here, these are considered the transitions from single-phase liquid to bubbly or dispersed bubble flow, followed by the transition from either bubbly or dispersed bubble to slug flow.

NETool Simulator Model

Well Case No. 2 from Chierici *et al.* [6] was modeled in the nodal analysis software NETool simulator, version 5000.0.4.2, with an additional development feature (not yet released) that allows the temperature to be set at every point along the wellbore. The multiphase model LMK-1 was used. LMK-1 has the same regime identification routine as shown in Figure 1 and Figure 2, based on modified Barnea regime transition functions [3].

A vertical well with an inner casing/liner diameter of 5.0 inches was created. A surface roughness of 5.0×10^{-4} inches (0.0127 mm) was assumed. The bottom 31 m (101.8 ft) of the wellbore were open to the reservoir, which was set to a pressure of 39.00 MPa (5,656.5 psi). The reservoir pressure and length of the inflow section are somewhat arbitrary, as matches to GOR and water cut were subsequently achieved through adjustment of the oil and water saturation in the reservoir and the well productivity index (PI) multiplier for each thermal model, as discussed below. The oil flow rate and tubing pressure at the wellhead were matched to the conditions given in Chierici *et al.* [6].

The thermophysical properties were set as follows. The oil API and specific gravity of gas were set to the reported values of 8.3 and 1.705, respectively. The surface tension of liquid oil in contact with gas was assumed to be a constant 0.03 N/m. The correlations used for the remaining properties are given in Table 1. A regression was performed to the given values of P_B, R_S, B_0, and μ_0 at the bottom-hole temperature. The mole fractions of H_2S, CO_2, and N_2 were assumed to be zero. The salinity was taken as 1.0 g per kg of water.

Table 1. Correlations for thermophysical properties of oil and gas.

property	**correlation**
dissolved gas/oil ratio	Standing
saturated oil formation factor	Standing
undersaturated oil formation factor	Vazquez-Beggs
saturated oil viscosity	Beggs-Robinson
undersaturated oil viscosity	Standing, Beal
gas volume factor	Standing-Katz
gas viscosity	Carr-Dempsey
gas viscosity at atmospheric pressure	Standing

Three thermal models for the wellbore were attempted. First, a linear temperature profile from the field-measured bottom-hole temperature to the wellhead temperature was applied. Then, the average temperature of 83°C (181.4°F) was applied to all downhole locations. Finally, the bottom-hole temperature of 109°C (228.2°F) was applied to all downhole locations.

Once the thermal model was selected, the saturations and well PI multiplier were adjusted until the production rates of oil, gas, and water were correctly replicated to within 0.5%. For example, for the case in which the average temperature was used throughout the wellbore, an oil saturation of 0.999457, a water saturation of 5.43×10^{-4}, no gas saturation, and a well PI multiplier of 12.804 produced the correct production rates of oil, gas, and water.

The predicted bottom-hole flowing pressure P_{BHF} was then compared to the field-measured value of 4,763.3 psi. The case with the linear temperature profile matched best, underpredicting P_{BHF} by only 1.6%. Using the average temperature throughout yielded an underpredicted P_{BHF} by 1.9%. Not surprisingly, the worst prediction was that in which the maximum temperature was used throughout, which resulted in an underprediction of P_{BHF} by 5.1%. Surprisingly, from simply considering P_{BHF}, it appeared that all thermal models performed well. However, when the conditions at each downhole location along the wellbore are considered, the differences are much more apparent. The details are artificially masked by matching the production rates.

Predictions of Wellbore Conditions

The liquid density and hydrostatic pressure are given versus measured depth in the wellbore for the three thermal models in Figure 3. The discontinuities in the liquid density curves mark the locations where the pressure drops below the bubble point pressure and dissolved gas begins to come out of solution. The local percent differences in hydrostatic pressure are as high as nearly 7 and 13% for the average and bottom-hole temperature cases compared to assuming a linear temperature profile.

Furthermore, the onset of the single-phase liquid to bubbly regime transition shifted by almost 150 m between the case with a linear temperature profile model and that of the bottom-hole temperature applied throughout. This is a significant difference from a design consideration, particularly for the placement of artificial lift. The second transition from bubbly to slug is also shifted and varies from measured depths of 76.5, 112, and 147 m for the linear, average, and bottom-hole temperature profiles, respectively.

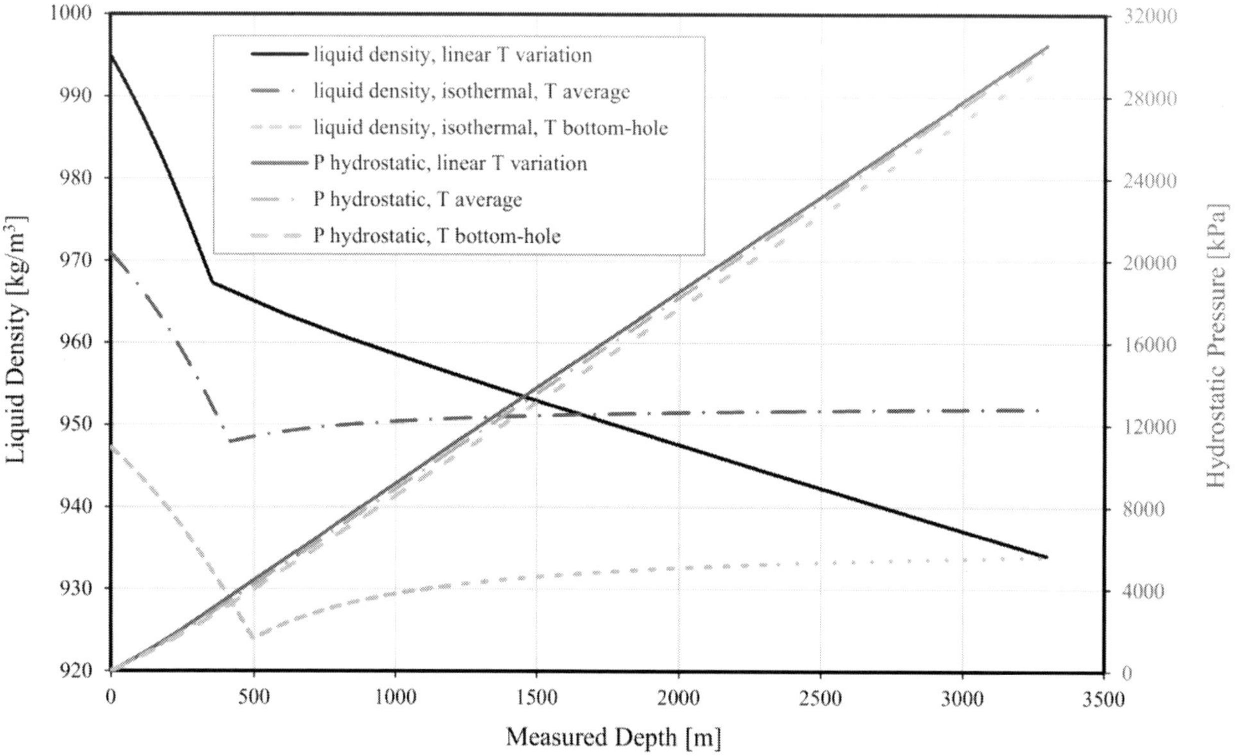

Figure 3. Liquid density and hydrostatic pressure along the wellbore for three thermal models.

Conclusions

The variations of liquid viscosity and gas density with temperature play significant roles for some regime transition functions. In particular, for very viscous liquids in vertical-upward flow, the dispersed bubble and churn regimes are stable over much more of the v_{SG} - v_{SL} space. For high-density gas flows, the annular regime becomes increasingly stable and consumes most of the churn and much of the slug regions.

A case study was performed for a vertical, heavy oil, onshore well. While the bottom-hole flowing pressure can be predicted reasonably well with several different thermal assumptions, the properties and flow regimes at a given downhole location can vary much more dramatically. For example, depths of regime transitions can vary on the order of 100 m. Accurate prediction of these depths is crucial to the design of artificial lift systems. Thus, an accurate thermal model should always be considered.

References

1 Hill, T., Johnson, T. and Hacala-Nicol, V. (2010). Steady-state, and interrupted, production through a deep water black oil system. *7th North American Conf. Multiphase Tech.*

2. Hart, A. (2014). A review of technologies for transporting heavy crude oil and bitumen via pipelines. *J. Petrol. Explor. Prod. Tech.*, **4**: 327–336.

3. Barnea, D. (1987). A unified model for predicting flow-pattern transitions for the whole range of pipe inclinations. *Int. J. Multiphase Flow*, **13**(1): 1–12.

4. Gomez, L.E., Shoham, O., Schmidt, Z., Chokshi, R.N. and Northug, T. (2000). Unified mechanistic model for steady-state two-phase flow: Horizontal to vertical upward flow. *Society of Petroleum Engineers J.*, **5**(3): 339–350.

5. Shoham, O. (2006). *Mechanistic Modeling of Gas-Liquid Two-Phase Flow in Pipes.* Richardson, Texas: SPE.

6. Chierici, G.L., Ciucci, G.M. and Sclocchi, G. (1974). Two-phase vertical flow in oil wells-Prediction of pressure drop. *J. Petro. Tech.*, **26**(8): 927–938.

7. Taitel, Y., Barnea, D. and Dukler, A.E. (1980). Modelling flow pattern transitions for steady upward gas-liquid flow in vertical tubes. *AIChE Journal*, **26**(3): 345–354.

8. Barnea, D., Shoham, O. and Taitel, Y. (1982). Flow pattern transition for vertical downward two phase flow. *Chemical Eng. Sci.*, **37**(5): 741–744.

9. Houchens, B.C., Popa, F. and Filippov, A. (2015). Sensitivity of dispersed-bubble flow regime identification over a broad parameter space and to various closure relations for mechanistic models of gas-liquid pipe flow. *WIT Transactions on Engineering Sciences*, **89**.

10. Brauner, N. and Barnea, D. (1986). Slug/churn transition in upward gas-liquid flow. *Chemical Engineering. Sci.*, **41**(1): 159-163.

11. Lin, P.Y. and Hanratty, T.J. (1986). Prediction of the initiation of slugs with linear stability theory. *Int. J. Multiphase Flow*, **12**(1): 79–98.

12. Barnea, D. and Taitel, Y. (1993). Kelvin-Helmholtz stability criteria for stratified flow: Viscous versus non-viscous (inviscid) approaches. *Int. J. Multiphase Flow*, **19**(4): 639–649.

13. Barnea, D. (1991). On the effect of viscosity on stability of stratified gas-liquid flow – Application to flow pattern transition at various pipe inclinations. *Chem. Eng. Sci.*, **46**(8): 2123–2131.

14. Taitel, Y. and Dukler, A.E. (1976). A model for predicting flow regime transitions in horizontal and near horizontal gas-liquid flow. *AIChE Journal*, **22**(1): 47–55.

444423 Numerical Modeling of Cavitation Erosion of Stagnation Point Flow

Abstract:

Cavitation damage caused by bubble collapse in the vicinity of solid surfaces, called cavitation erosion, is a complex process. It involves complicated flow phenomena combined with the reaction of the solid surface material. Several models have been proposed to predict cavitation erosion in hydrodynamic equipment, but limited number can be coupled with CFD. In this work, we take a look at one of these models, namely Dular's model, and apply it to a stagnating flow that undergoes cavitation. Several parameters will be investigated, 2D vs. 3D effects, turbulence model selection, and steady vs. transient. Comparison with experimental data will be presented.

Copyright © American Institute of Chemical Engineers. All rights reserved.

442150 Dynamic Ultrasound Phase Separation Measurements: From Liquid/Liquid to Gas/Liquid Separation

Abstract:

The performance of separators is crucial to oil field operation. Multiphase flow occurs in separators at low velocity compared to pipes and gravity forces cannot be neglected. To understand the behavior of such system, we have developed a technique based on ultrasound applied for liquid/liquid separation. The main advantage of this technique vs. well known bottle test is to provide more parameters than the conventional evolution of settled water volume such as the height and location of the dense pack zone, water droplet sedimentation kinetic or the evolution of water fraction in the separated oil. Experimental set-up equipped with dynamic ultrasound scanners of different scale and operating modes were used to develop a methodology to select processes, determine operating conditions and chemical treatment strategy to meet the BS&W specification for several projects.

The systems generated in the presence of gas and liquids in separators can also lead to complex interfaces that require advanced measurement technology, especially for foaming which is a well known problem in oil & gas separation.

After an introduction on the use of the ultrasound methods on liquid/liquid separation, the purpose of this work is to present the results of ultrasound methods applied to gas/liquid separation and to compare with standard foaming test based on visual observations and image analysis. Preliminary results have shown that foam decay and degassing from liquid occurring simultaneously are adequately captured on a system where air is dispersed in a stock tank oil. Results from real hydrocarbon system after depressurization of a cell equipped with dynamic ultrasound scanner will also be presented. Similarly to liquid/liquid separation, the perspectives of using ultrasonic tools to improve gas/liquid separation processes and chemical treatment requirement will also be also detailed.

Copyright © American Institute of Chemical Engineers. All rights reserved.

444663 Assessment of Liquid Distribution in a VGO-Hydroprocessing Trickled Bed Reactor with Computational Fluid Dynamics

Abstract:

Assessment of liquid distribution in a VGO-hydroprocessing trickled bed reactor with computational fluid dynamics

J. J Arias B., jjariasb@unal.edu.co

A. Molina, amolinao@unal.edu.co

CFD simulations of the liquid distribution in a Trickled Bed Reactor (TBR) hydro-processing Vacuum Gas Oil (VGO) were carried out. TBRs are vertical columns with fixed-bed reactors within. Inside the reactor liquid and gaseous phases pass through the bed on a concurrent, down-flow mode. This kind of reactor has the capacity to host different kind of catalytic reactions. Frequently they are used to refine crude oil cuts, as is the case in hydro-processing, where VGO reacts with hydrogen to eliminate heteroatoms by hydrogen substitution (hydro-treating) and saturate broken bonds generated by catalytic cracking with hydrogen atoms (HydroCracking (HC). Uniform liquid distribution at the top of this kind of reactors, while fundamental to guarantee proper operation, is very difficult to achieve.

To understand the effect of liquid flow mal-distribution at the top of TBR, four sections of an industrial-size HC reactor were modeled. The first section involves the surroundings of the gas-liquid distributor, itself the second section. A third simulation considered the distribution tray. The bed was the fourth section modeled considering a reaction mechanism particularly designed for the analysis of HC processes. To simulate each of these sections ANSYS Fluent was used. Sections two and three were modeling using a VOF model while the fourth section used an Eulerian model. The meshes used were built on ICEM with cell numbers not below 200000 with non-uniform sizes.

The simulations showed that the velocity profile at inlet of the gas-liquid distributor unit has velocities values larger than 15 m/s. In the gas-liquid distributor there is an expansion of the gas while flows through it, also the gas-liquid distributor disperses the liquid across the distributor's cross-section. The effect of liquid distribution in the reactor performance did not show a significant coupling between injection points. A maldistribution factor was used to evaluate the dispersion of the liquid at the top of the bed and at the outlet compartment of the gas-liquid distributor, this factor was used like a measurement of the improvement of the liquid distribution. Finally the catalytic bed simulation shows that there is an inefficient use of the catalyst bed and therefore a lower performance of the reactor when a non-uniform liquid profile is used.

Volume liquid fraction profiles, velocity profiles at different location and kinetic performance of the TBR were analyzed on each of the simulations to enhance the liquid distribution over the TBR.

Copyright © American Institute of Chemical Engineers. All rights reserved.

443857 Distinguishing and Prediction Periodic and Chaotic Multiphase Flow in T-Junction Microchannel Obtained from Optical Time-Series Signals

Abstract:

The stability and regularity of multiphase flow in T-junction microchannel were investigated in this study. In virtue of homemade optical measuring system, time series related to two-phase flow dynamics under different operating conditions, fluids and channel lengths were collected and analyzed through a nonlinear characteristic parameter of Kolmogorov entropy (K_E). The results reveal that higher K_E corresponds to irregular and unstable flow behavior, while the lower K_E refers to desirable operating condition with steady and orderly flow behavior. Meanwhile, most of the Taylor flow regime appeared at proper operating conditions with small K_E presented steady flow behavior. In order to further understand and estimate K_E, a concise equation derived from definition of K_E was proposed, which included bubble break-up impact and flow rate ratio of gas and liquid. Results of the calculations and the experiments were in good agreement. The derived equation possesses effectiveness to identify and ensure stable and predictable multiphase flow in microchannel.

Copyright © American Institute of Chemical Engineers. All rights reserved.

Recognition and primary test on secondary flow in T-type outlet

Mengda Jia[1], Chaoyu Yan[1], Jianfei Song[1], Yaodong Wei[1], Huimin Zhang[1], and Faqi Zhou[1], (1)College of Chemical Engineering, China University of Petroleum, Beijing, China

Abstract

This work presents an recognition and primary test on alternative secondary flow in T-type outlet. The phenomenon in experiment is shown in pictures. The numerical study is performed by Fluent CFD, in witch a pure gas flow and gas-particle flow are simulated in a riser with a T-type outlet. Euler-Euler two-fluid model and RNG k-ε model was adopted. In the results, the basic flow field in T-type outlet possessed a pair of Dean vortex. In a certain condition, there existed a pair of switching vortex, witch alternative occupied the main space of T-type outlet. The changing process of the switching vortex along with time was discussed. When gas velocity was 10 m/s, gas flow state cycle time was 0.16 s. Affected by the delay effect from particle, gas-solid flow showed a lower frequency in switching than pure gas flow. And the flow states cycle time increased with the particle volume friction grow.

Keywords: secondary flow; T-type outflow; multiple-flow; riser

1 Introduction

T-type exit is one kind of exit in riser reactor, and air cushion forming in its blind pipe can slow the erosion on wall caused by up moving particle. Therefore, T-type exit become one of the most widely used exit of riser. In order to know the time-average characteristics of T-type exit as a part of riser, Most study were focused on its strong constraints and the pressure drop between its inlet and outlet[1].Ulrike[2] and Chian[3] experimentally studied two-phase flow in riser with a T-type structure exit, and the result showed that there exist drastic particle back-mixing in T-type exit, and it play a significant role in affecting the distribution of pressure and particle concentration in upper part of riser with blow the exit. Considering the effect of T-type exit, Cheng[4] defined it as an exit of strong constraints. Comparing with other exit configurations (C-type and L-type)Zheng[5], flow field in T-type exit possess unique feature. Engelandt[6] compared Four kinds of T-type exits(they called T abrupt outlet) with different sizes, and including 3 kinds extension height and 2 kinds outlet pipe diameter. The results indicated that, there exist a vortex in T-type exit. With different configuration size, such as decreasing extension height, the position of the vortex shifts downwards and towards the riser wall opposite the riser outlet. De Whide[7] performed a series of numerical simulation, and finally they found the unstable flow in T-type exit.

Dean firstly discovered and defined secondary flow in cross section of round exit pipe. Many studies measured flow in exit pipe using apparatuses such as laser Doppler velocimeter, and physical parameters as time-average velocity, instantaneous velocity and pressure distribution were systematic investigated. The flow at the exit of pipe exhibits a pair of symmetrically oscillatory vortexes, and they alternatively occupy the flow field with time, so-called "swirl-switching".[8~13].After then, the "swirl-switching" phenomenon aroused interest of many researchers, and they performed experimental[14] and numerical[15]study on it. In the experimental studies literature, by the difficulty of measurement, the result can not describe the whole flow field, even the actual motion of the symmetrically oscillatory vortexes. Tunstall and Harvey[14] consider the switching motion is resulted of turbulent flow and abrupt configuration. But the following research denied these opinion since "swirl-switching" phenomenon was observed in a mild curvatures as well[10]. Nevertheless, these kind of research were only limited to quarter exit, and the phenomenon of symmetrically oscillatory vortexes were not observed in other types of pipe.

For T-type exit, study on its internal secondary flow was not given. Therefore, this study is focused on simulating the flow in T-type exit.

2 Numerical study

2.1 RNG k-ε modeling Euler-Euler two-fluid model and RNG k-ε model were adopted to simulate the gas-solid flow in T-type outlet. Gidaspow drag model was used to simulate the interaction between gas and particle. Finite volume method was adopted to spatial dispersion and uncoupled implicit scheme to solution computation. Second-order upwind difference scheme was adopted to momentum and diffusion, and SIMPLE scheme to coupling the pressure and velocity. There are three faces bounding the calculation domain : the inlet, the outlet, and the wall. The specified gas velocity at the inlet are imposed in original flow field computation. The outlet boundary condition is set up as an outflow boundary. In the region near, the no slip boundary condition is applied. particle diameter is 67μm, particle density is 1500 kg/m-3. Inlet gas velocity is 9.2m/s and particle velocity is 1.46m/s

2.2 Numerical procedure Fig 1 give a sketch of the geometry and mesh grid of the T-type exit witch consists of 3000 mm long upstream riser and 800 mm long outlet pipe.The T-type exit was of 240 mm in diameter and 80 mm in extent pipe, and the diameters of riser and outlet pipe are 240 mm and 160 mm respectively. The grid independency is tested using computational grid between 100000 and 500000 cells, and the computational grid of approximately 390000 cells shows optimum performance. To fully investigate the flow in special T-type structure, 2 cross-section was chosen to be analyzed.

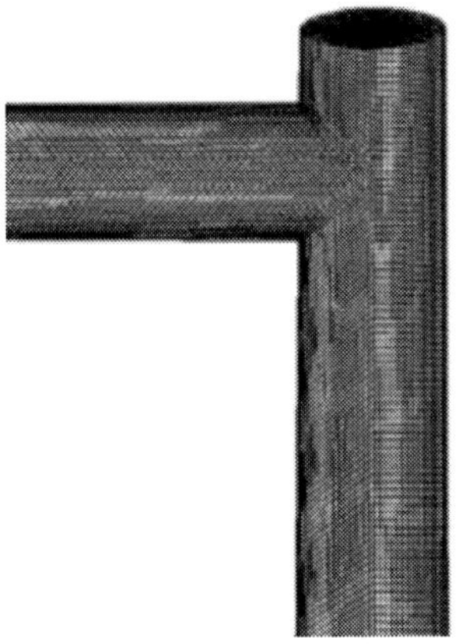

Fig.1. Geometry and mesh grid of T-type outlet

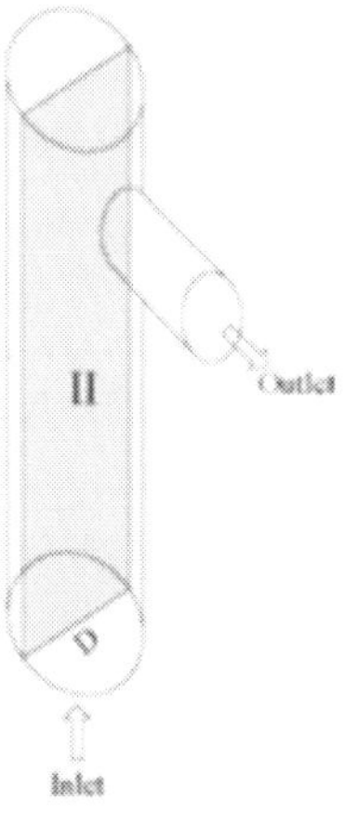

Fig.2. Cross-section to study

3 Recognition in experiment

In riser, when particles are transported by gas flow to the outlet, affected by outlet restraint, a mount of particle reach the top of riser and back down, forming a mass of particle agglomeration. Start from the top region of riser, particle back-flow spread down along the riser, and particle steamers can been clearly observed on the wall(bai et al.1991a). The back-flow phenomenon is show in Fig.3.

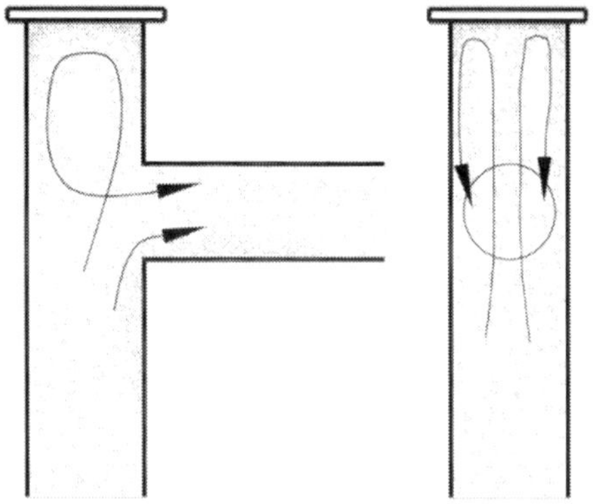

Fig.3. particle flow in T-type outlet

In recent experiments, a phenomenon of unstable back-flow occur in T-type outlet. Fig4(a)~Fig4(e) showed the continuous change of particle flow in T-type outlet, and the angle of view is shown as Fig.1. In Fig.4(a), a apparent white line in the right of riser is the particle steamer, witch flow downward. Then particle steamer change to a parabola in Fig4(b), and the position move to the center of riser. In Fig.4(c), the particle steamer move to left, and back to right in the same way in Fig.4(d)~Fig4(e). Generally, the particle back-flow swing in T-type outlet with the times, witch regular turn to left and right alternatively. Up to now, these is the first report about the swing flow in T-type exit.

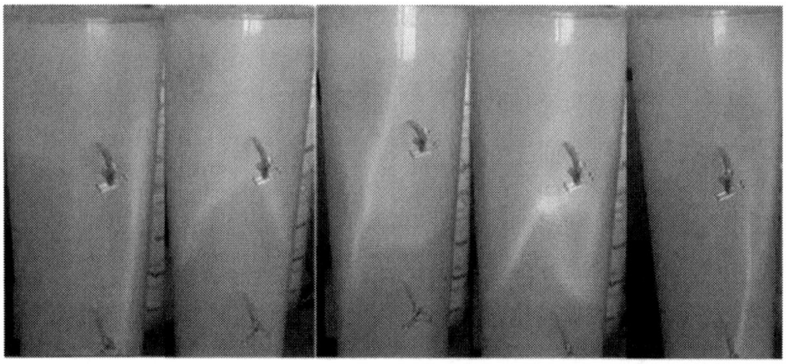

(a) (b) (c) (d) (e)

Fig.4. continuous change of particle flow in T-type outlet

4 gas flow simulation

4.1 basic flow field Fig.5 show streamline chart and velocity contour of cross-section I, in the condition of v_g=10 m/s. In Fig.5, gas enter the extend pipe and emerge to the center region of the pipe, and retrace downward to each side after reaching the top, resulting to two symmetrical vortexes formed. The size of these two vortexes is bigger than the one in cross section I. The symmetrical vortexes in T-type exit have a similar structure with the dean secondary flow in 90° bend. In 90° bend, fluid move fast in center, while the flow near wall is significant affected by boundary layer. An imbalance is caused between centrifugal force and pressure gradient, so flow velocity decrease with nearness to the boundary layer, giving rise to a vortex. The situation in T-type exit has a little different, as in Fig.5. While gas flow into the exit, local pressure increase due to existence of the roof, and gas flow along with negative pressure gradient. Under the combined affection of boundary layer, the gas near the wall can't keep a upstream flow, then a velocity contrast appears between flow near wall and in center, resulting to a vortex. In addition, the flow in T-type exit mainly tend to the exit pipe, so the vortexes is spiral.

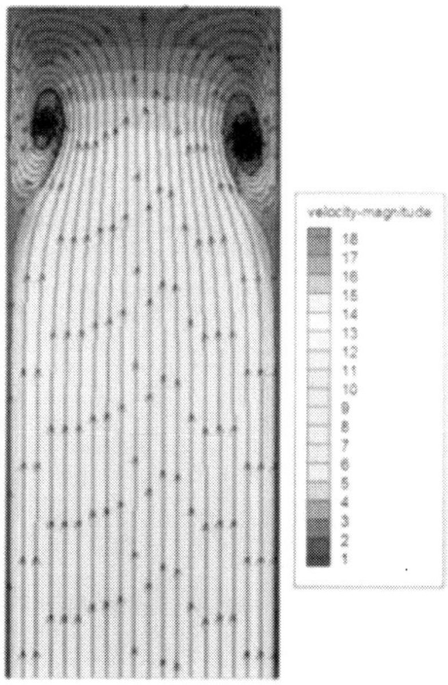

Fig.5. streamline and velocity distribution in cross-section I

4.2 evolution of switching vortexes In previous discussion, three vortexes is obviously discovered in T-type exit flow field. On the basis of this, the distribution of streamline in cross-section I were investigated at different moments.

Fig.6 show flow field of cross-section I at different moments, these six pictures are typical flow regime which can clearly describe flow in T-type exit. Fig.6(a) is specified as 0 moment, and Fig.6(b)~Fig.6(h) are the following moments, such as t_a=0 s, t_b=0.04 s, t_c=0.07 s, t_d=0.10 s, t_e=0.13 s, t_g=0.16 s . In cross-section I, firstly(moment a) two vortexes locate in a symmetrical position, and they are similar in size. Then(moment b) the left vortex shrink to a smaller one and its center move to wall and a under position. Meanwhile, the streamline between these two vortexes slightly turn to left. The left shrunken vortex return back to the position at moment a and became bigger again, and the streamline between the vortexes. At moment d, the right vortex shrink in the same way with the left one in moment b, and the center streamline turn to right. In moment e and f, the flow field return to the state of moment a and b.

Thus a conclusion can be get. Under a certain Reynolds number, there exist a pair of swing symmetrical vortexes in cross-section I, and the vortexes present a cycle regularity.

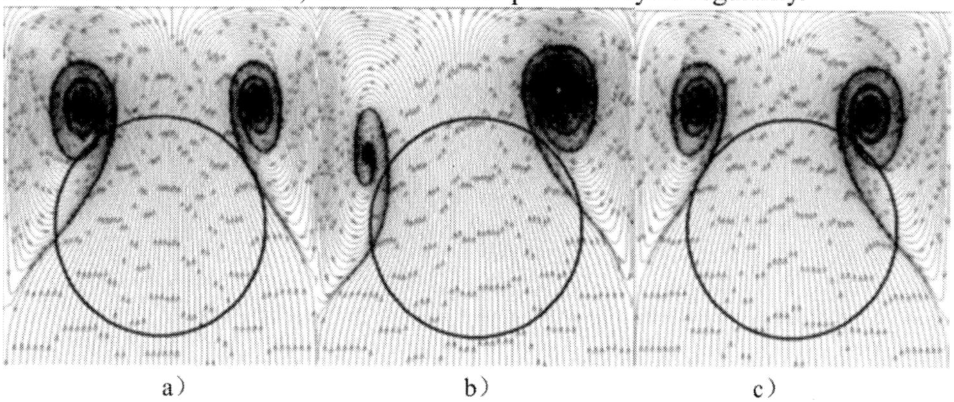

a) b) c)

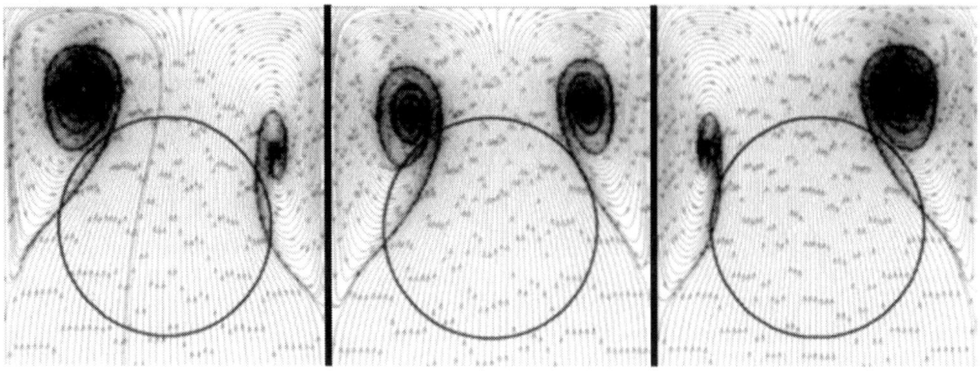

d) e) f)

Fig.6. distribution of streamline in cross-section I in different moment

5 multi-phase simulation

Secondary flow affect the flow behavior of particles. In same gas and particle velocity, 3 volume friction was chosen, such as 0.01(Gs=138 kg/m2·s), 0.03(Gs=414 kg/m2·s), 0.08(Gs=1104 kg/m2·s).

Fig.7 show change of particle velocity vector with the time when the particle volume friction was 0.01. When t=4.1 s, back-flowed particles are accumulated on the right wall. Then, main particle move to the center(t=4.3 s). Particles concentrate upon the left wall of riser(t=4.4 s), and return right when t=4.6 s. Fig.7 and Fig.7 show change of particle velocity vector with the time when the particle volume friction was 0.03 and 0.08 respectively. Particle flow in Fig.7 and Fig.7 show a same behavior with Fig.7, but the flow state cycle time is different.

In conclusion, flow states cycle time is 0.6, 0.9 and 1.1 when the particle volume friction was 0.01, 0.03 and 0.08. It mean that, a higher particle volume friction lead to a long flow state cycle time. On the basic of the gas flow simulation result, the vortexes alternative time is 0.16 s. Ignoring the calculation error, the flow states cycle time of gas-solid flow is much higher than the one of pure gas. These means, the source of the swing flow is the gas secondary flow, so-called the alternatively swing symmetrical vortexes, and the exist of particles delay the vibration of the gas. When the particle volume friction increase, the delay effect is grow.

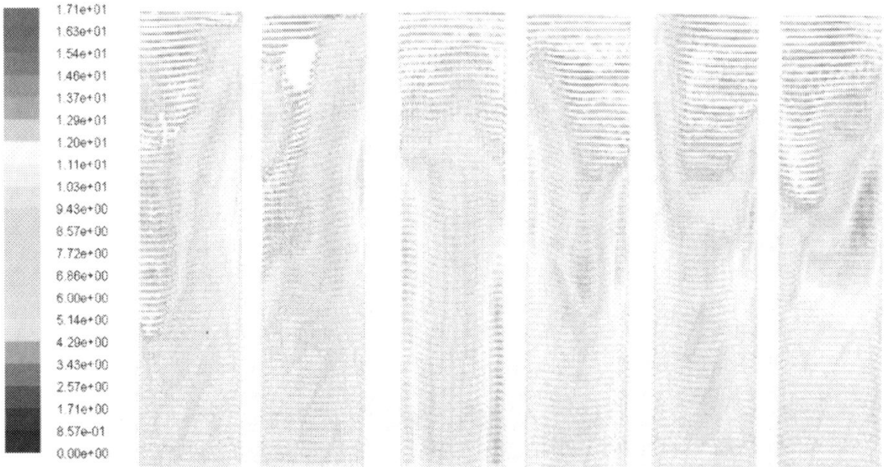

t=4.1s t=4..2s t=4.3s t=4.4s t=4.5s t=4.6s

(a)

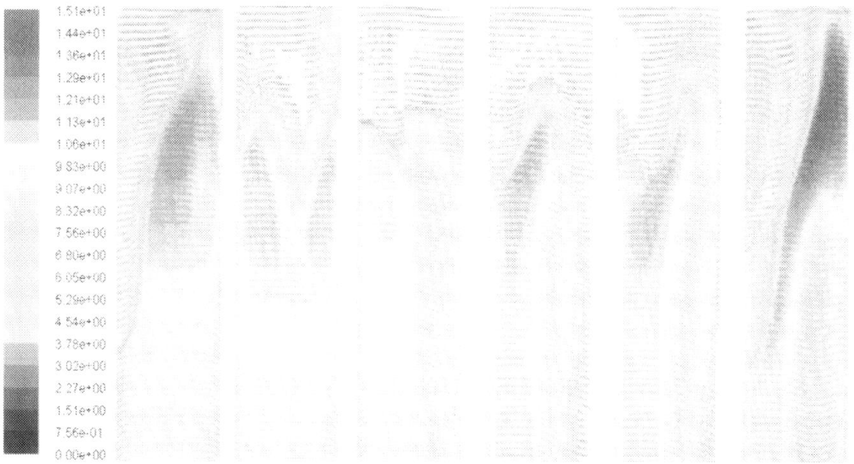

t=4.7s t=4.9s t=5.0s t=5.2s t=5.3s t=5.6s

(b)

t=5.5 s t=5.7 s t=6.0 s t=6.1s t=6.3 s t=6.6 s

(c)

Fig.7 Vectors of particle velocity vs.time in the cross-section I

Conclusion

The goal of this work is to recognize and primarily test the secondary flow in T-type outlet, by interesting the regular experimental phenomenon. A numerical approach was developed to help understanding flow behavior. FLUENT CFD soft ware was used as a basis for numerical simulations and necessary user function were added to adapt the software to the physical problem in hand. Firstly a pure gas flow was simulated in a certain gas velocity, and a pair of switching vortex is found, witch alternative occupy the main space of T-type outlet. The changing process of the switching vortex along with time is discussed and gas flow state cycle time is 0.16 s. Then a series gas-solid flow simulation was performed, and the result showed a lower frequency in switching than pure gas flow. Affected by the delay effect of particle, the flow states cycle time increase with the particle volume friction grow.

Reference

[1] Juray De Wilde, Guy B. Marin, Geraldine J. Heynderickx. The effects of abrupt T-outlets in a riser: 3D simulation using the kinetic theory of granular flow. Chemical Engineering Science 58 (2003) 877~885
[2] Ulrike Lackermeier, Joachim Werther. Flow phenomena in the exit zone of a circulating fluidized bed. Chemical Engineering and Processing. 41 (2002) 771-783
[3] ChianWen Chan, Anke Brems, Shiva Mahmoudi, et al. Joseph Gargiuli. PEPT study of particle motion for different riser exit geometries. Particuology.2010. 8(6):623–630
[4] Yi Cheng, Fei Wei, Guoqiang Yang, Yong Jin. Inlet and outlet effects on flow patterns in gas-solid risers. Powder Technology, 1998 (98) 151-156
[5] Zheng, Q., Jin, Y., Zhang, H., 1995. Effect of bed exit geometry (end effect) on hydrodynamic behaviour of gas-solid flow in CFB combustor. Proc. Fluidiza- tion VIII, 461–483.
[6] G. Vanengelandt,G.J.Heynderickx n, JurayDeWilde 1, G.B.Marin. 2001, Experimental and computational study of T- and L-outlet effects in dilute riser flow. Chemical Engineering Science.1995:66:5024-5044
[7] De Wilde, J., Marin, G.B., Heynderickx, G.J., 2003a. The effects of abrupt T-outlets in a riser: 3D simulation using the kinetic theory of granular flow. Chem. Eng. Sci. 58, 877–885.
[8] Sudo, K., Sumida, M., Hibara, H., 1998. Experimental investigation on turbulent flow in a circular-sectioned 90˚ bend. Exp. Fluids 25, 42 - 49.
[9] Sudo, K., Sumida, M., Hibara, H., 2000. Experimental investigation on turbulent flow in a circular-sectioned 180˚ bend. Exp. Fluids 28, 51 - 57.
[10] Rütten, F., Schröder, W., Meinke, M., 2005. Large-eddy simulation of low frequency oscillations of the Dean vortices in turbulent pipe bend flows. Phys. Fluids 17, 035107.
[11] Sakakibara, J., Sonobe, R., Goto, H., Tezuka, H., Tada, H., Tezuka, K., 2010. Stereo-PIV study of turbulent flow downstream of a bend in a round pipe. In: 14th Int. Symp. Flow Vis., 21 - 24 June, EXCO Daegu, Korea, pp. 1 - 11.
[12] Sakakibara, J., Machida, N., 2012. Measurement of turbulent flow upstream and downstream of a circular pipe bend. Phys. Fluids 24, 041702.
[13] Kalpakli, A., Orlu., Alfredsson, P.H., 2012. Dean vortices in turbulent flows: rocking or rolling? J. Vis. 15, 37 - 38.
[14] Tunstall, M.J., Harvey, J.K., 1968. On the effect of a sharp bend in a fully developed turbulent pipe-flow. J. Fluid Mech. 34, 595 - 608.
[15] Hellstrue, F., 2010. Numerical computations of the unsteady flow in turbochargers. PhD thesis, KTH Mechanics, Stockholm, Sweden.
[16] Bai D, Jin Y, and Yu Z(1991a). Cluster observation in a two-dimensional fast fluidized bed, fluidization'91-Science and Technology(edited by kwank, M and Hasatani, M) pp.110-115, Science press, Beijing

444941 To Insulate or Not: A Perspective Based on Gelation

Abstract:

Gelation of waxy oils during a shutdown can result in difficulties restarting the flow. This is due to the fact that the yield stress of the gel needs to be overcome for restart. The yield stress of a particular oil is dependent on many factors such as the shear and thermal histories of the oil. A gel formed under shear, for example, will be weaker than a gel formed under static conditions. In this presentation, we look at this effect and how it affects the decision of whether to insulate a line or not.

Copyright © American Institute of Chemical Engineers. All rights reserved.

444985 Investigation of Inhibitors Efficacy in Wax Deposition Mitigation Using a Laboratory Scale Flow Loop

Abstract:

The deposition of wax in the inner walls of subsea crude oil pipelines is severe operational issue in the production and transportation of oil. Chemical inhibition technique is one of the promising method to mitigate the risk associated with wax deposition in the pipelines. The efficiency of different polymer based chemical inhibitors (PI-A and PI-B) on wax deposition at laminar and turbulent conditions is investigated using a laboratory scale flow loop using Shah Deniz (SD) condensate. The thickness calculated by the deposit mass collected is consistent with that obtained by pressure drop method. For the laminar flow, the deposit thickness has significantly decreased in the presence of the wax inhibitors. After 72 h, the thickness reduced 35 % in PI-A and 80% PI-B. The wax content in the presence of PI – B (38%) is significantly higher compared to PI-A (11%) and NO PI (11%). In turbulent flow, the deposit thickness decreases more than 2.5 times compare to laminar flow with and without inhibitors. After the 72 h, the thickness reduced 45 % in PI-A and 85% in PI-B. The wax content in the presence of PI – B (40%) and PI-A (25%) increases significantly compared to NO PI (18%). The paraffin inhibition efficiency was calculated based on the wax mass. PI-B is more effective in overall wax mass reduction compared to PI-A in both laminar and turbulent flow.

Copyright © American Institute of Chemical Engineers. All rights reserved.

444870 Nano-Particle Based Waxs Inhibitors for Environmentally Friendly Pour Point Depressing in Waxy Crude

Abstract:

Crystallization of paraffin waxes complicates the low temperature flow properties of waxy crudes, generating problems in pipeline transportation. An efficient solution to this problem is using pour-point depressants (PPD), such as polymeric additives, altering the crystallization of the paraffin waxes and hence the rheological properties since these are highly dependent on the paraffin crystal morphology. Recently the combination of nano-materials with traditional polymeric PPDs have shown promise [1, 2].

This work combines inert nano-particles with polymeric PPDs to greatly enhance the effectiveness of these additives on a model waxy crude system. The effectiveness and mode of operation of these hybrid nano-particles are discussed, discussing the effect of polymer coverage of the nano-particles, and how the environmental difference in using hybrid nano-particles in comparison to neat chemicals.

1. Yang, F., et al., *Hydrophilic Nanoparticles Facilitate Wax Inhibition*. Energy & Fuels, 2015. **29**(3): p. 1368-1374.

2. Yang, F., et al., *Polymeric Wax Inhibitors and Pour Point Depressants for Waxy Crude Oils: A Critical Review*. Journal of Dispersion Science and Technology, 2015. **36**(2): p. 213-225.

Copyright © American Institute of Chemical Engineers. All rights reserved.

445003 Development of a Probabilistic Approach to Assessment of the Restartability of Waxy Crude Pipeline

Abstract:

Pipeline restart failure due to oil gelation after prolonged shutdown is the main threat of the safe operation of waxy crude pipeline. Many factors influencing the restartability of a waxy crude pipeline such as oil properties, operation and environmental parameters have uncertainty. Current methods for pipeline restartability assessment take no consideration of this uncertainty, including the traditional force balance equation, which is known to be over-conservative, and the numerical method based on unsteady heat transfer and thixotropic fluid flow.

Reliability-based method is an effective approach to solve this kind of uncertainty problems, which has been widely used in some other fields such as structure engineering. In the pipeline industry, ISO issued the first standard of reliability-based limit state method for the design and assessment of oil and gas pipeline structure in 2006, i.e. ISO 16708. In this study, we introduce this reliability-based approach to solve the problem of pipeline restartability with consideration of parameter uncertainty, and the restartability is reported in the way of restart failure probability.

A limit state function for restart failure is developed based on a model describing the unsteady flow and heat transfer of pipeline during the shutdown and restart process, and an algorithm is developed which can solve the probability model fast and accurately, which is necessary to break through the bottle-neck in computation of the failure probability with truly-unsteady flow and heat transfer model because of the huge computation load. And moreover, the target safety level, i.e. the maximum acceptable failure probability, for the restart of crude pipelines is proposed. Consequently, quantitative assessment of restart safety of waxy crude pipeline can be achieved in the form of restart probability, and the lowest allowable pumping temperature and the maximum allowable time for pipeline shutdown can be determined based on the reliability of pipeline restart.

Compared to the traditional methods which determine the pumping temperature of waxy crude by mainly referring to its pour point and determine the restartability of a waxy crude pipeline without taking the uncertainty of the influencing parameters into consideration, this probabilistic approach provides new and scientific methodology for safe and economic operation of waxy crude pipelines.

Copyright © American Institute of Chemical Engineers. All rights reserved.

444418 Wax Deposition Modeling of Oil-Water Flows Using CFD

Abstract:

Wax deposition in subsea oil pipelines is a significant economic problem in the oil and gas industry. The deposition of wax particulates on the pipe wall can block the pipe and cause significant losses in oil production. Thus, it is necessary to accurately predict wax deposition possibility in oil transport pipelines. Wax deposition involves multiphase flow undergoing mass and heat transfer. In this work, we model wax deposition by solving transient Navier-Stokes equations in a non-isothermal oil-water flow in pipelines. Wax gel is modeled a porous media, where the velocity and pressure drop are calculated from combined Darcy and Navier-Stokes equations. Numerical results are compared to published experimental work.

Copyright © American Institute of Chemical Engineers. All rights reserved.

444066 Studying the Effect of Stripping Light Hydrocarbons from Shale Oil Using Model Oil System

Abstract:

Wax deposition is a common and costly problem in oil industry, especially in pipelines, storage tanks, barges, and containers. Wax precipitation occurs when the temperature in the tanks fall below the cloud point or wax appearance temperature (WAT). WAT is dependent on many factors such as the rate of cooling, composition of the oil, cooling mechanism (quiescent or sheared cooling). Oils derived from shale formations often have both high vapor pressure and significant wax concentrations. Vapor pressure standards for transportation of crude oil requires the lighter ends stripped from oil. In this work, the effect of stripping light hydrocarbons on the WAT and the rheological property will be presented using a model-oil system. The lighter hydrocarbon used are pentane, hexane and heptane. We determine the WAT, gelation temperature, yielding behavior for different shear histories, and creep recovery using a cone and plate system for the model oil system with and without the lighter hydrocarbons. In addition, polarized light microscopy is used to visually probe into the size and aspect ratio of the wax crystals formed at different conditions. A correlation is drawn between the stripping of lighter hydrocarbons and the change in WAT and other rheological properties.

Copyright © American Institute of Chemical Engineers. All rights reserved.

2016 AIChE Spring Meeting
2016-04-13

Delaying hydrates formation and preventing hydrates agglomeration by adding MEG to formation water containing various concentration of NaCl

Hyunho Kim (hyunho.kim@snu.ac.kr)
Advisor : Yutaek Seo

Dept. of Naval Architecture and Ocean Engineering
Seoul National University

Contents

Introduction

Introduction

- **Gas Hydrate**
  - Safety problems and costs associated with hydrate blockages
  - Network of cages of **water molecules** that can trap small paraffin guest (**gas**) molecules
  - Generated at low T and high P

    → **Injecting hydrate inhibitor**

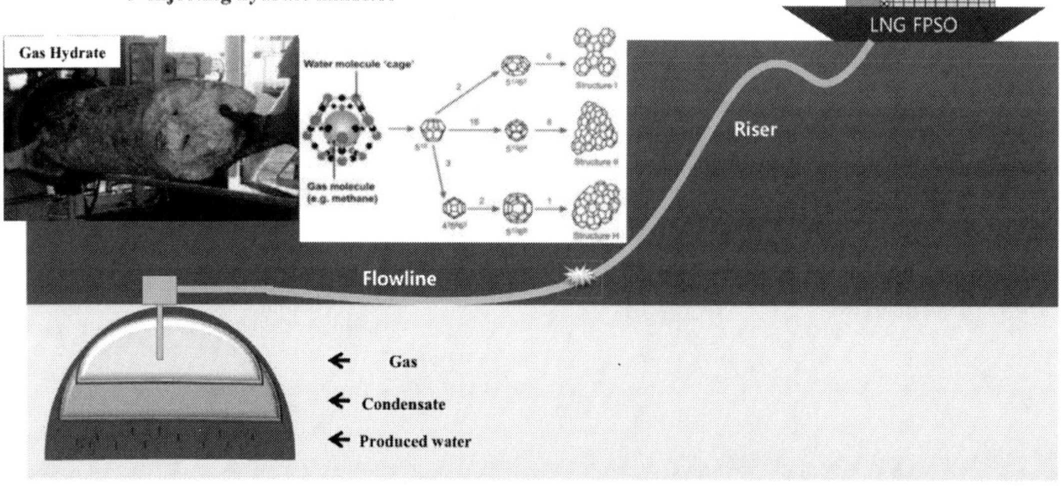

Introduction

➢ Hydrate inhibitors

Thermodynamic Inhibitor (THI)	Low Dosage Hydrate Inhibitor (LDHI)

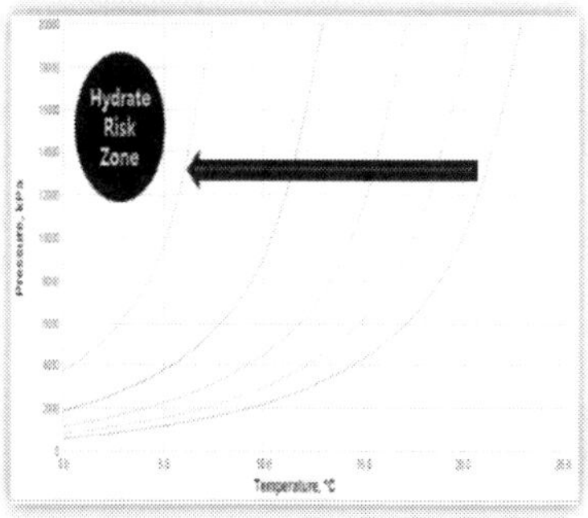

1) Kinetic Inhibitor (KHI) : To delay hydrate form
2) Anti-agglomerant (AA) : To allow hydrate
No plug formation by emulsifying water droplet

< Shifting hydrate phase equilibrium ,THI>

- Shifting hydrate phase equilibrium

Introduction

➢ Under-inhibition concept

- To control effectively using **less amount** of THI than required

Thermodynamic Inhibitor (THI)	+	Low Dosage Hydrate Inhibitor (LDHI)
Large volume injection (20 ~ 60 wt%) 1) Monoethylene glycol (MEG) 2) Salts (NaCl)		Small volume injection (0.5~2.0wt%) 1) Kinetic Inhibitor (KHI) 2) Anti-agglomerant (AA)

⬇

Reduction in the amount of THI injection

Introduction

➢ Thermodynamic inhibitor

MEG	Salts (NaCl) in produced water
• Expensive (10 times more than Methanol)	• Salty water produced along with the oil and gas (up to 25 wt%) Ex> Hibernia : 4.6 ~ 19.5 wt% • 90 % of the salt = NaCl

- To use inhibitor ?
- **Under-inhibition** ?

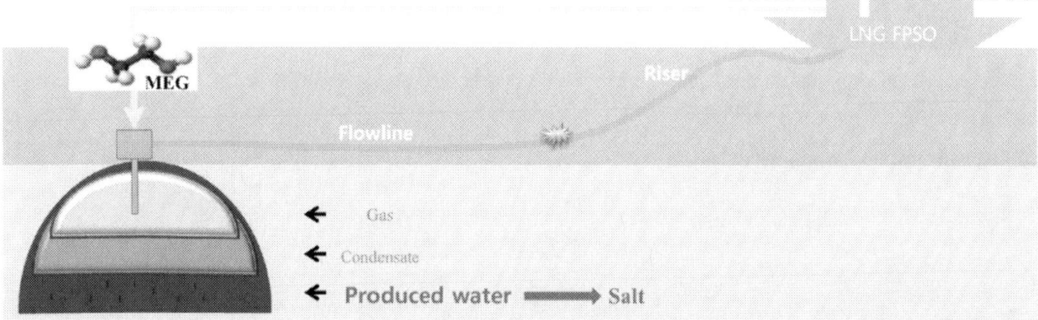

Objectives

Objectives

Salts (NaCl) in **produced water** + Monoethylene glycol (**MEG**)

- To suggest **MEG under-inhibition** considering the influx of **produced water**
  - To investigate **synergistic inhibition effect** of MEG and NaCl with various salinity in MEG

Experiments

Experiments

➢ Experiment apparatus

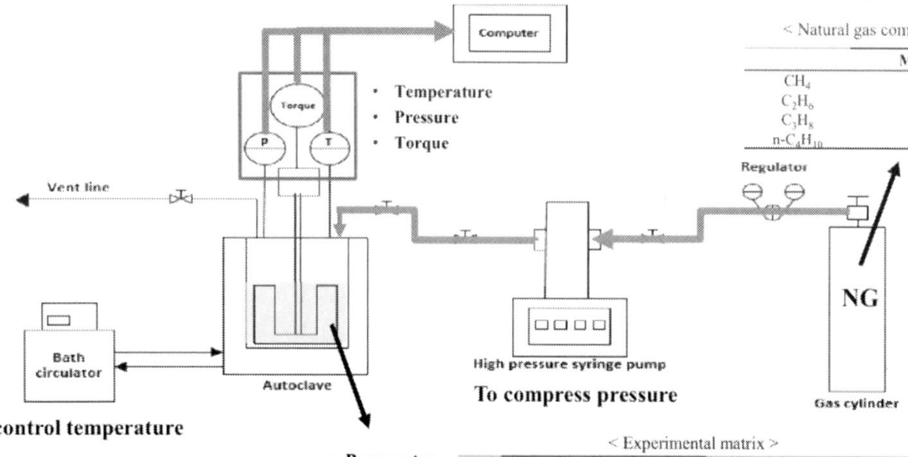

< Natural gas composition>

	Mol %
CH_4	90
C_2H_6	6
C_3H_8	3
$n\text{-}C_4H_{10}$	2

< Experimental matrix >

	MEG Concentration (wt%)	NaCl Concentration (wt%)
1	-	-
2	20	-
3	20	3.5
4	20	10

Experiments (2)

2) Kinetic performance

➢ To observe kinetics of hydrate formation

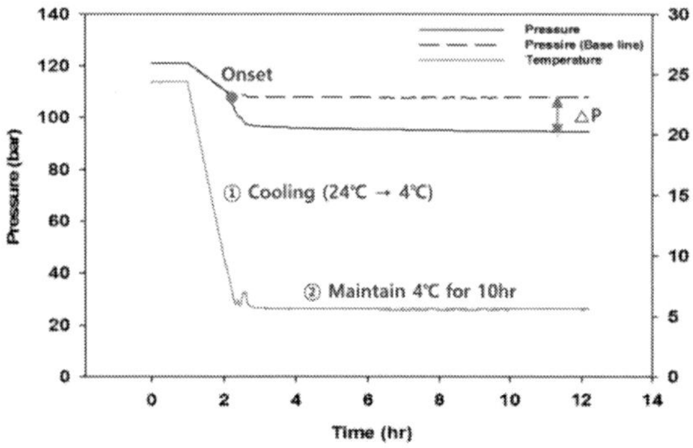

< P and T changes during the experiment>

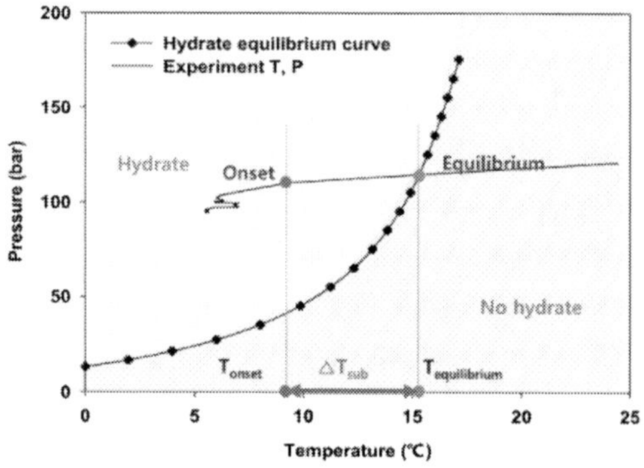

< Hydrate phase equilibrium curve >

➢ Indicator of kinetic performance

1) Torque: The degrees of agglomeration and deposition of hydrate particles

2) Subcooling temperature (ΔT_{sub}): Driving force forming hydrate

Results and discussion

Results and discussion (2)

2) Kinetic performance (1/3)

- **The change of torque is decreased with MEG + NaCl.**
  - Although the hydrate formed, the torque change keeps constant in (MEG + NaCl).
    - (MEG + NaCl) prevents the rapid agglomeration.

The degrees of agglomeration

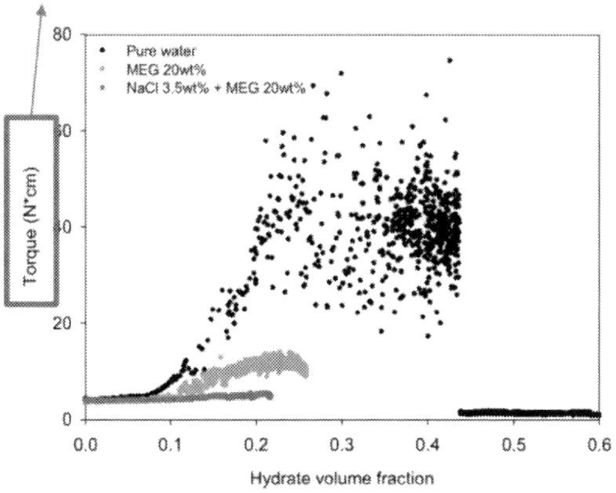

< Torque changes during hydrate formation>

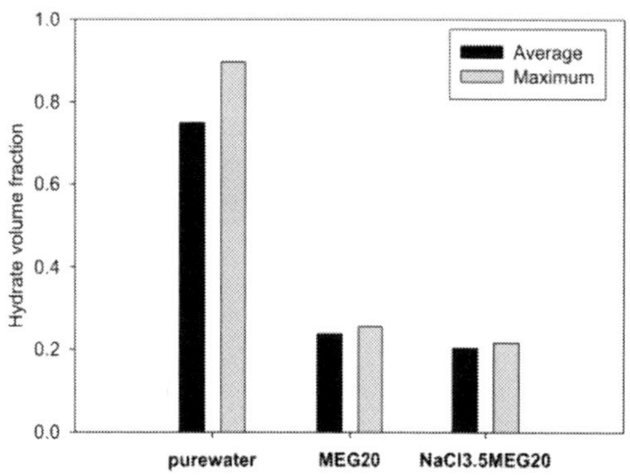

< Hydrate volume fraction>

Results and discussion (2)

2) Kinetic performance (2/2)

- The initial hydrate growth rate is decreased with (MEG + NaCl).

System	r_{ini} (mmol/min)
Pure water	43.27
NaCl 3.5wt%	10.59
MEG 20wt%	22.52
NaCl 3.5wt% + MEG 20wt%	4.69
NaCl 10wt%+ MEG 20wt%	-

4.8 times↓

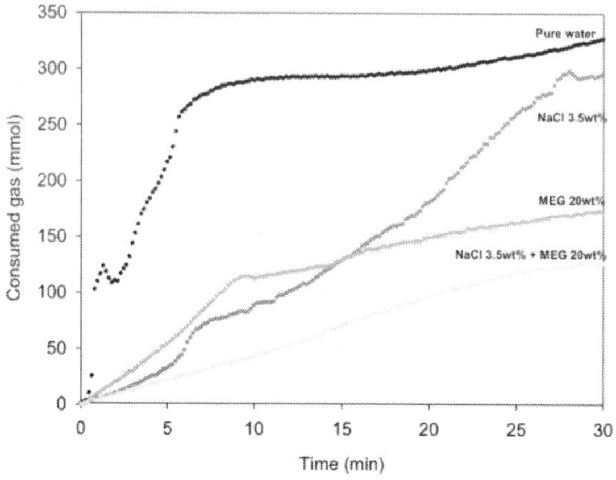

<Gas consumption profiles after hydrate onset time>

Conclusions

Conclusions

- **(MEG + NaCl) solution had synergistic kinetic inhibiting performance.**
  - **The torque change kept constant in (MEG + NaCl).**
  - The growth rate at NaCl 3.5 wt% + MEG 20 wt% was 4.8 times slower than the rate at only MEG 20 wt%.
- **It is possible to under-inhibit MEG considering the influx of produced water.**

ENGINEERING
COLLEGE OF ENGINEERING
SEOUL NATIONAL UNIVERSITY

17

Acknowledgement

- This research was supported by a grant from the project of development of methodology for subsea production system configuration and verification funded by the Ministry of Trade, Industry and Energy Republic of Korea.
- This research was also supported by the Industrial Strategic Technology Development Program (Development of flow assurance and organic acid/calcium removal process for the production of offshore opportunity crude) of Korea Evaluation Institute of Industrial Technology funded by the Ministry of Trade, industry & Energy (MI, Korea).

ENGINEERING
COLLEGE OF ENGINEERING
SEOUL NATIONAL UNIVERSITY

444391 Risk-Based Estimation of Hydrate Blockage Risk in Oil and Gas Flowlines

Abstract:

The numerical simulation of gas hydrate formation in oil and gas flowlines has traditionally employed the coupling between advanced hydrate reaction models with a multiphase transportability framework. In recent years, the Hydrate Flow Assurance Simulation Tool (HyFAST) has been developed as an open framework, where laboratory-based hydrate growth and transportability models are paired with a simplified hydrodynamic model to describe the frictional pressure loss due to hydrate growth in multiphase flow. This presentation introduces the third generation of the tool, which incorporates a critical new feature: the frictional pressure loss due to hydrate slurry formation or wall deposition is coupled within the momentum balance across the flowline. This accomplishment means that HyFAST 3 is able to estimate the risk of hydrate blockage based on flowline velocity approaching zero. To our knowledge of the field, HyFAST 3 is the first global flowline simulation tool to connect hydrate-based frictional pressure loss to flow velocity; as a consequence, the tool does not require rules of thumb to assign a risk category to the amount of hydrate in the system. This presentation will demonstrate how this novel risk assessment method may be deployed in the industry, which may be coupled with laboratory characterisation to determine the maximum allowable hydrate fraction in the system without increasing the risk of hydrate blockage.

Copyright © American Institute of Chemical Engineers. All rights reserved.

444378 Anti-Agglomerant Benchmarks at Interfacial and Macroscopic Length Scales

Abstract:

Clathrate hydrates are crystalline inclusion compounds which can form in subsea flowlines and lead to the formation of blockages. To mitigate this risk, thermodynamic hydrate inhibitors (THI), such as monoethylene glycol, are added to shift the pipeline operating pressures out of the hydrate stability region. As the production of oil and gas moves into deepwater, the required length of subsea tiebacks will increase significantly. In such operating scenarios, there is a increase in the driving force for hydrate formation, and more THI is required to fully suppress hydrate formation. To maintain the viability of deepwater assets, alternative hydrate remediation technologies must be benchmarked and validated at the benchtop and pilot scale. In the present work, we have investigated and benchmarked industrial anti-agglomerant (AA) chemicals using a micromechanical force apparatus. AAs are surfactants that target the hydrate particle surface, resulting in a non-aggregating and non-depositing hydrate-in-oil slurry. This work presents the first quantitative measurements of AA adsorption on to the hydrate interface for both first and current-generation industrial AAs. We observe that the current generation AA is a more powerful surfactant, but may be further improved by increasing its selectivity for the hydrate-oil interface. The cohesive force results were up-scaled using a high pressure rheometer, which demonstrated that the AA provided a beneficial reduction in slurry viscosity and strongly affected the hydrate crystal growth rate.

Copyright © American Institute of Chemical Engineers. All rights reserved.

444384 Hydrate Formation and Deposition Rate Studies in a Gas-Dominant Flowloop

Abstract:

Gas hydrate formation and blockage is a critical engineering concern for the oil and gas industry. While many studies over the past two decades have focused on quantifying hydrate formation and slurry behaviour in oil-dominant systems, there is a dearth of information available with which to assess hydrate growth rate or blockage severity in gas-continuous systems where liquid water may exist in either annular or stratified flow. This investigation deploys a single-pass flowloop in Perth, Western Australia, which was designed to simulate a gas-dominant flow at laboratory scale. The flowloop has been deployed to study the formation and deposition of gas hydrate in systems with variable liquid holdup (1-10 vol%) and subcooling up to 20 K. The results demonstrate that, at high gas velocities, approximately 50% of the hydrate formed exists as a deposit on the wall; the remaining hydrate particles are distributed through the gas phase in flow. While the total hydrate formation rate increased directly with subcooling, the rate of deposition and wall build-up depended heavily on flow regime. The frequency and severity of sloughing were observed to increase with subcooling, where visual analysis from the sapphire window was used to estimate average slough size. The results provide critical input parameters to aid in the development of gas-dominant formation and deposition models within hydrodynamic simulation engines.

Copyright © American Institute of Chemical Engineers. All rights reserved.

444394 Novel Experimental Methods to Quantify Surfactant Adsorption to Hydrophilic-Lipophilic Interfaces

Abstract:

In conventional energy production, the formation of water- and hydrate-in-oil dispersions generates a strong degree of frictional pressure loss, and may enforce sizing requirements for downstream receiving facilities. Depending on the nature of the system's chemistry, surfactants may be injected into the production system to manipulate or minimise this free energy and improve the ability of the system to flow. Conventionally, the performance of these surfactants is validated qualitatively through macroscopic benchtop methods, including rocking cells. In this study, we have deployed three experimental methods to quantify the competition between surfactants at hydrophilic-hydrophobic interfaces. First, pendant drop interfacial tensiometry (IFT) was adapted to quantify a matrix of adsorbed surfactant densities via Gibbs isotherm analysis, for surfactants exposed to water-oil interfaces over ten orders of magnitude in concentration. Second, micromechanical force (MMF) tests were deployed over a similar concentration range to quantify the adsorption of surfactants at the clathrate hydrate-oil interface. Third, Raman spectroscopy methods were adapted to quantify the extent of surfactant adsorption at the gas hydrate-oil interface above the dissolved concentration in the continuous phase. Comparison of IFT with MMF/Raman data provide an attractive means by which to characterise the hydrate-philicity of surfactant molecules, where the size of the hydrophilic group is a critical differentiating factor; the results demonstrate that these experimental methods may be used to tune a new generation of hydrate-selective surfactant chemicals.

Copyright © American Institute of Chemical Engineers. All rights reserved.

444395 Magnetic Resonance Measurements of Hydrate Shell Growth in Oil

Abstract:

In crude oil systems, hydrate may nucleate and grow at the interface of emulsified water droplets. When the droplets are large, initial hydrate nucleation may extend through only a portion of the shell, maintaining a liquid water core that has been suggested to grow slowly over time. In this work, we present novel benchtop nuclear magnetic resonance (NMR) pulsed field gradient (PFG) and relaxation measurements to monitor the formation of a hydrate shell from large water particles dispersed in a continuous cyclopentane phase. These techniques allow us to monitor hydrate growth inside the opaque exterior shell, revealing novel information about the evolution of shell morphology over time. NMR relaxation measurements were primarily used to monitor the hydrate shell growth kinetics, while PFG NMR diffusion experiments were used to determine the nominal droplet size distribution of the unconverted water core. A comparison of the mean droplet sizes obtained directly via PFG NMR and independently deduced from relaxation measurements revealed that the assumption of a shell model for hydrate systems was correct, but only after approximately 24 hours of shell growth. Initially, the hydrate growth was observed to be limited by heat transport, resulting in a porous shell with a larger surface area than a sphere of equivalent volume. After 24 hours, continued hydrate growth became mass transfer-limited, and the shell became thicker, more spherical, and less porous.

Copyright © American Institute of Chemical Engineers. All rights reserved.

444402 Rheological Characterisation of Hydrate-in-Oil Slurries

Abstract:

Gas hydrates are ice-like solids that readily form in oil and gas flowlines under high pressure and at low temperature. In oil-dominant systems, the formation of a viscous hydrate-in-oil slurry has been identified as a primary mechanism contributing to hydrate blockage, but there is a dearth of data available to inform hydrate slurry viscosity models. This work deploys a temperature-controlled, high-pressure rheometer with a vane blade rotor to study the rheological properties of methane hydrate-in-crude oil slurries, which are reacted from water-in-oil emulsions. The work represents one of the first applications of a high-pressure vane blade controlled-stress rheometer, which has enabled quantitative comparison of slurry viscosity as a function of the hydrate volume fraction. After the hydrate reaction in each experiment, flow curves are captured for the steady-state hydrate-in-oil slurries and the slurry yield stress is measured after a prolonged shut-in period. The results demonstrate that hydrate-in-oil slurries are typically shear thinning, and the apparent viscosity of the system decreases as temperature increases. The flow curves are compared with current industry-leading models for the hydrate slurry, which deviate up to 50% from the measured values.

Copyright © American Institute of Chemical Engineers. All rights reserved.

443678 Cyclopentane Hydrate Formation in Water-in-Model Waxy Oil Emulsions

Abstract:

Mitigating flow assurance issues, such as plugging of flowlines due to hydrates and wax, is of great importance in deepwater and subsea production environments due to safety and environmental concerns, as well as high operating costs. Extensive research has been conducted on each of these flow assurance issues; however, there is limited understanding on the interaction of hydrates and wax simultaneously. The lack of fundamental understanding of hydrate formation in waxy oil emulsions motivates our current experimental work in such systems. For this work, a visualization technique is employed to investigate hydrate formation in water-in-cyclopentane and waxy oil emulsions. An Olympus BX53 polarized optical microscope with shear cell and temperature control (-50°C to 450°C) stage is used to determine rate of hydrate formation, quantify droplet size distribution, hydrate crystals and morphology. Cyclopentane hydrate formation is quantified in a paraffin containing model oil (Crystal Plus 70T) system stabilized using either solid particles or surfactants.

Copyright © American Institute of Chemical Engineers. All rights reserved.

16AIChE
Spring Meeting, Houston, TX

The Role of Deasphalted Oil in the Stability of Asphaltenes

Abhishek Punase
Texas A&M University
Petroleum Engineering Department, Texas A&M University, 3116 TAMU College Station, Texas 77843, USA
abhishekpunase@tamu.edu

Berna Hascakir
Texas A&M University
Petroleum Engineering Department, Texas A&M University, 3116 TAMU College Station, Texas 77843, USA
hascakir@tamu.edu

[Insert here any copyright claims you wish to assert]

Prepared for Presentation at
American Institute of Chemical Engineers
2016 Spring Meeting
11th Global Congress on Process Safety
Houston, Texas
April 10-14, 2016

UNPUBLISHED

AIChE shall not be responsible for statements or opinions contained
in papers or printed in its publications

The Role of Deasphalted Oil in the Stability of Asphaltenes

Abhishek Punase
Texas A&M University

Berna Hascakir
Texas A&M University
Petroleum Engineering Department, Texas A&M University, 3116 TAMU College Station, Texas 77843, USA
hascakir@tamu.edu

Keywords: Asphaltenes stability, Phase envelope, SARA fractions

Abstract

Asphaltenes represent the heaviest fraction of the crude oil. The physical and chemical variations during oil production destabilize the asphaltene molecules and cause asphaltene precipitation. This work deals with the analysis of the change in asphaltene phase behavior due to the compositional variations. Two crude oil samples which have different asphaltene content, API gravity, and viscosity were examined. The stability of asphaltenes was investigated by defining each crude oil with a four-component system. Hence, the crude oils were divided into their Saturates, Aromatics, Resins, and Asphaltenes (SARA) fractions by following a standard method which uses n-pentane to separate asphaltenes. First, the spectral characterization on the crude oils and on their SARA fractions was achieved by Fourier Transform InfraRed (FTIR) spectroscopy. Indirect assessment of the polarity of the crude oil and its fractions were also carried out using dielectric constant measurements. Then, ternary phase diagrams were constructed by using the blending of saturated (n-pentane) and aromatic (toluene) hydrocarbons with asphaltenes fraction at varying amounts. The three-component two-phase (solid-liquid) diagrams have been constructed based on experimental data. Asphaltenes are soluble in aromatic solvents and insoluble in normal alkanes, thus, while the increase in amount of n-pentane decreases the asphaltene stability, the increase in toluene reestablishes the stabilization. While the systematic analyses conducted in this study enhance our understanding, it is necessary to observe behavior of asphaltenes after the interaction with crude oils own saturates and aromatics fractions which would provide better insights towards the role of deasphalted crude oil fractions in maintaining the stability of asphaltenes.

1. Introduction

Asphaltenes represents the heaviest and most polar fraction of the crude oil and its precipitation is known to cause severe flow assurance related issues [1, 2]. Changes in thermodynamic conditions within production and transportation flow-lines can lead to compositional variation of crude oil and reduce the stability of asphaltenes molecules [3, 4, 5, 6]. Understanding the role of deasphalted oil fractions in maintaining asphaltenes stability can provide important insights towards the precipitation mechanism and improve the efficiency of different inhibition processes. Deasphalted oil, also known as maltenes, is a mixture of saturates, aromatics, and resins fractions of the crude oil [7]. Molecules containing straight, branched, or cyclic alkane chains are considered to be part of saturates fraction. Aromatics portion of the crude oil comprises of aromatic hydrocarbons with trace amounts of heteroatoms like nitrogen, sulfur, and oxygen [8, 9, 10]. Molecular and compositional behavior of resins is similar to asphaltenes, which can contain complex aliphatic hydrocarbons attached to condensed aromatic and naphthenic rings containing heteroatoms [11, 12, 13, 14]. The polar resin fraction is known to peptize asphaltenes molecules and enhance its stability [1]. On the other hand, interaction of asphaltenes with saturates or aromatics fractions requires more clarity as the presence of different impurities within these non-polar fractions can impart varying partial polarity and affect asphatlene stability [15, 16]. This study focusses on investigating the role of deasphalted oil in maintaining the overall stability of asphaltenes. Two crude oil samples, originating from North and South America and having distinct physical properties were analyzed, and two-phase ternary diagrams representing the stability of asphaltenes fractions were developed.

2. Experimental Procedure

Physical properties like density and viscosity of two crude oil samples were measured using Anton Paar DMA 4100 density meter and Brookfield DV-III Ultra rheometer, respectively. Both crude oils were then separated into its Saturates, Aromatics, Resins, and Asphaltenes (SARA) fractions by following ASTM D2007-11 method [17]. Asphaltenes fraction obtained through the fractionation method represents n-pentane insoluble portion of the crude oil. Compositional analysis of crude oils and its individual fractions were carried out using Fourier Transform InfraRed (FTIR) spectroscopy technique. Analysis of FTIR spectra is useful in estimating the chemical composition and molecular structure of the test sample and gives an indirect assessment of polarity. To comprehend the role of deasphalted oil fractions on asphaltenes stability, systematic control experiments were conducted by blending asphaltenes with saturated (n-pentane) and aromatic (toluene) hydrocarbons at different weight ratios. Results and observations from the control experiments were used to generate two-phase (solid-liquid) ternary diagrams.

3. Experimental Results and Discussion

3.1 Physical Properties and SARA fractionation

Physical characterization of the two crude oil samples analyzed in this study was conducted through density and viscosity measurements (**Table 1**).

Table 1 – Physical properties and SARA fractions of crude oil samples. API gravity is obtained for standard conditions (60 °F) while all the other measurements corresponds to room temperature (22.3 °C).

Crude Oil	Gravity (°API)	Viscosity (cP)	Saturates (wt%)	Aromatics (wt%)	Resins (wt%)	Asphaltenes (wt%)
Oil1	17.12	496	30.03	41.84	15.56	12.57
Oil2	10.01	19,200,000*	11.05	30.47	16.06	42.41

*** Viscosity value is obtained through viscosity-temperature correlations due to limitations of the equipment. Similarly API gravity values of both the crude oil samples were estimated using density-temperature correlations to obtain value at standard conditions. Relative error ranges for the measurements are ±0.15%, ±1%, ±15% for API gravity, viscosity, and SARA fractionation, respectively.**

Table 1 highlights a wide deviance in the physical properties of the bulk oil samples with the density varying from 17.12 to 10.01° API and viscosity from 496 to 19,200,000 cP for Oil1 and Oil2, respectively. SARA fractionation results also indicate large compositional variation between the crude oil samples. Analysis of physical properties and SARA fractions can provide useful information about the bulk oil samples, but understanding the stability of asphaltenes requires further investigation into its chemical composition and interaction with other crude oil fractions.

3.2 Compositional Analysis

Elemental composition and bond structure of the molecule plays an important role in establishing its stability and interaction propensity. Thus, carbon, hydrogen, and heteroatoms content of the crude oils and their corresponding asphaltenes fractions were analyzed and the results are given in **Table 2**.

Table 2 – Elemental composition of crude oil and their n-pentane insoluble asphaltene fractions [20].

Samples	Elemental composition of Crude Oils, wt%			Elemental composition of n-C5 Asphaltenes, wt%		
	C	H	Heteroatoms	C	H	Heteroatoms
Sample1	81.30	11.50	7.20	78.78	7.42	13.80
Sample2	80.80	10.30	8.90	78.35	7.63	14.02

***Heteroatom content is calculated by excluding the contribution of C and H from bulk weight (100%). It should be noted that C and H may have both organic and inorganic surface, hence, the portion of C and H may contribute to heteroatom content in the form of carbonates or bicarbonates.**

It can be observed that the heteroatoms present in crude oil shows higher affinity towards asphaltenes and remains concentrated in the asphaltenes fraction. Presence of heteroatoms in the asymmetric molecular structure of asphaltenes can increase its polarity and impact the stability. Polarity indicates the non-uniform distribution of charges in a molecule due to presence of elements having different electronegativity and asymmetric bond geometry [18, 19].

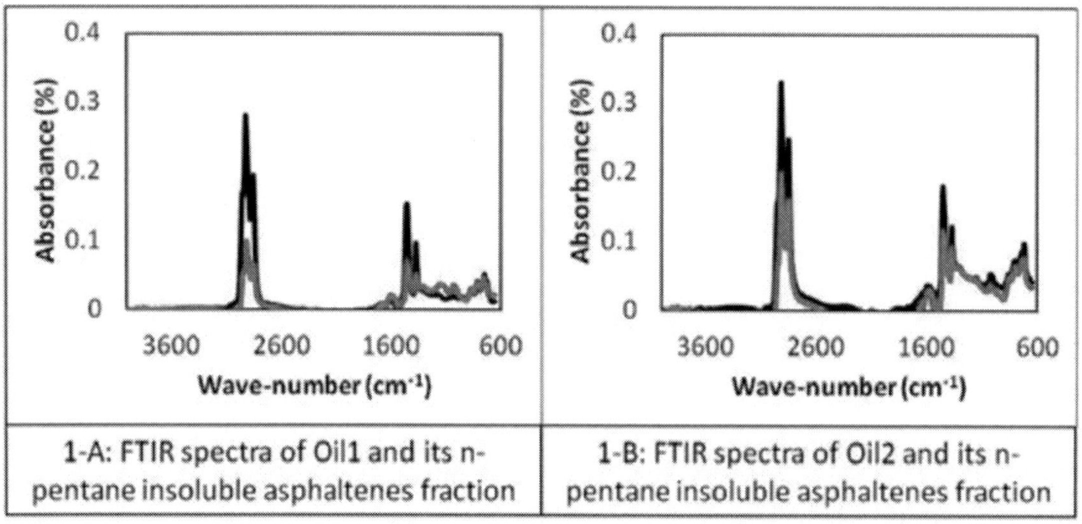

Figure 1: FTIR spectra of Oil1 and Oil2 (red curves) and their n-pentane insoluble asphaltenes fractions (black curves).

Analysis of the bond structure of crude oils and their SARA fractions were carried out using FTIR spectroscopic analysis. FTIR spectroscopy generates unique spectral signatures related to the absorption pattern of the material. These signatures at specific wavenumber region correspond to the frequency of vibrations between the bonds making up the atom of the test samples. The absorbance peaks corresponding to 2953-2853 cm^{-1} and 1458-1376 cm^{-1} wave-number represents bond stretch associated with $-CH_3$ and $-CH_2$ bends of a saturated aliphatic hydrocarbon, respectively. The absorption around 3065-3040 cm^{-1} with smaller peaks near 1700 cm^{-1} and 1600 cm^{-1} wavenumber region indicates aromatic structures containing C=C and C=O bonds, respectively [21]. The FTIR spectra of Oil1 and Oil2 together with their corresponding asphaltenes fractions are shown in **Figure 1**. It can be observed that the crude oils and asphaltenes spectra show signature peaks at similar wavenumbers with Oil2 having higher absorbance compared to Oil1 in the aliphatic hydrocarbon region. Similar trend is followed by the asphaltenes fraction, hence indicating that Oil2 asphaltenes fraction may be less polar than the asphaltenes fraction of Oil1. FTIR spectra of saturates, aromatics, and resins fractions of both crude oil samples are included in **Figure A-1**. In addition to FTIR analysis, an indirect assessment of crude oils and asphaltenes polarity through dielectric constant measurements was conducted and the results are shown in **Figure 2** [16]. Dielectric constant value is known to directly relate with the polarity of the test sample [18, 22, 23, 24, 25], therefore, the results presented in Figure 2 validates that crude oil and asphaltenes fractions of Oil1 are more polar than Oil2.

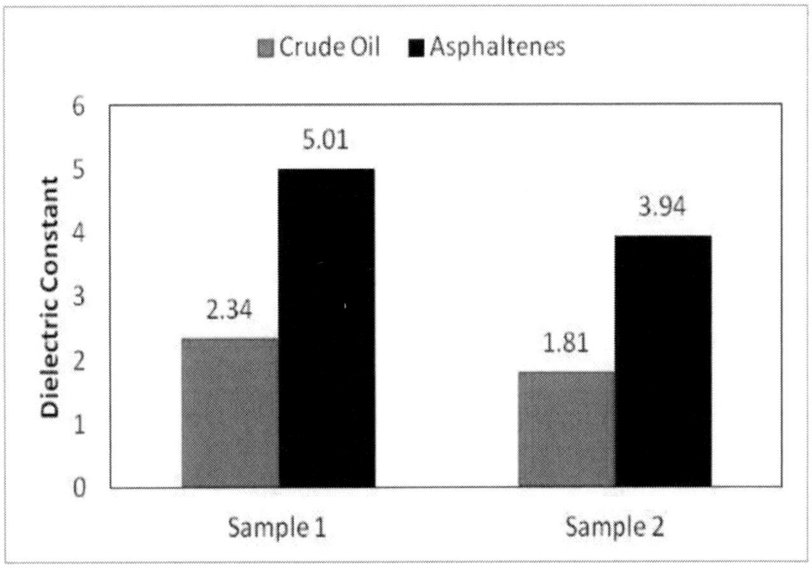

Figure 2: Dielectric constant of crude oils (red bars) and their n-pentane insoluble asphaltenes fractions (black bars) (Punase et al. 2016).

3.3 Ternary Phase Diagrams of Asphaltenes

Assessment of asphaltenes stability and its mutual interactions with other crude oil fractions is carried out by conducting systematic control experiments and developing the two-phase (solid-liquid) ternary diagrams. In this study, n-pentane and toluene were used as representative saturates and aromatics fractions of crude oil. Solvent solutions with varying pentane-toluene weight ratios were prepared and asphaltenes fraction was added to the mixture. Filtration of the three-component (pentane-toluene-asphaltenes) mixture through a vacuum filter assembly provided an estimate of solid fraction content. Solvent concentration resulting in retention of 50 wt% solid residue in the filtrate was considered as the solid-liquid phase boundary concentration. **Figure 3** presents the ternary phase diagram of both Oil1 and Oil2 asphaltenes fractions. The area occupied by the solid phase region of Oil1 asphaltenes is observed to be higher than Oil2 asphaltenes. This behavior agrees with the dielectric constant results and suggests that Oil1 asphaltenes may be more polar than the asphaltenes fraction of Oil2. Since the overall dielectric constant of Oil1 is higher than Oil2, it can provide a more stable medium for the asphaltenes fractions and may result in lower precipitation of asphaltenes.

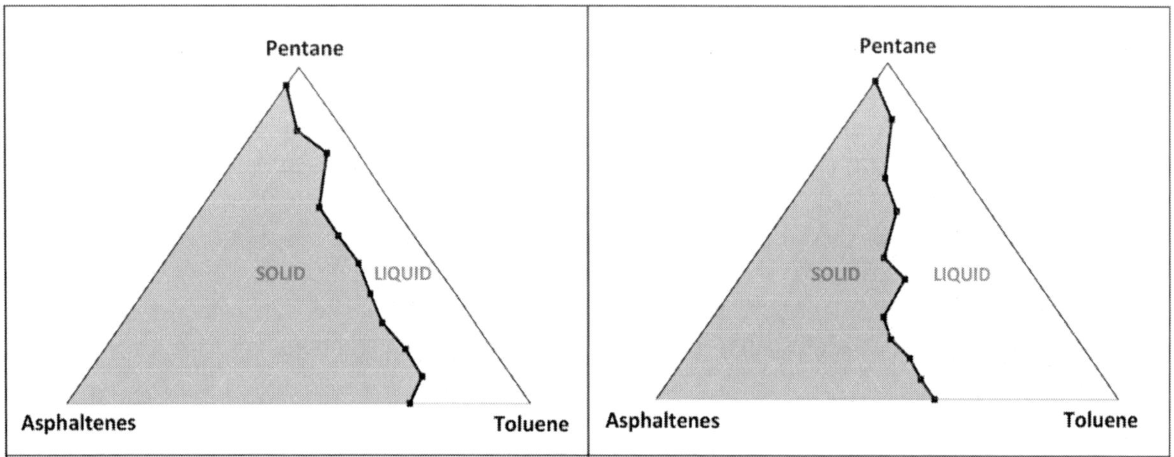

3-A: Ternary phase diagram between pentane, toluene, and asphaltenes fraction of Oil1

3-B: Ternary phase diagram between pentane, toluene, and asphaltenes fraction of Oil2

Figure 3: Ternary two-phase diagrams of Oil1 and Oil2 asphaltenes fractions.

4. Conclusions

Precipitation of asphaltenes within the flow lines can cause severe flow assurance problems and understanding the stability of asphaltenes fraction within crude oil can be useful in preventing the aggregation tendency. Two crude oil samples with distinct physical properties and asphaltenes content were analyzed. Elemental composition, FTIR spectroscopy, and dielectric constant measurements of crude oils and asphaltenes fraction indicated that the polarity of Oil1 is higher than Oil2. Interactions of asphaltenes fractions with saturated (n-pentane) and aromatic (toluene) hydrocarbons were investigated by developing two-phase (solid-liquid) ternary diagrams. Oil1 asphaltenes was observed to have lower affinity towards the solvents and occupied higher solid phase region as compared to the asphaltenes fraction of Oil2. The composition and overall polarity of crude oil is observed to play an important role in maintaining asphaltenes stability.

5. Acknowledgements

We acknowledge the Heavy Oil, Oil shales, Oil sands, & Carbonate Analysis and Recovery Methods (HOCAM) Research Team at Texas A&M University, Petroleum Engineering Department, for their help.

6. Nomenclature

SARA	Saturates, Aromatics, Resins, Asphaltenes
FTIR	Fourier Transform InfraRed
ASTM	American Society of Testing and Materials
API	American Petroleum Institute

7. References

[1] Speight, J.G., '*The Chemistry and Technology of Petroleum*", 2nd ed., New York City, Marcel Dekker, 1991

[2] Mullins O.C. and Sheu E.Y., "*Structures and Dynamics of Asphaltenes*", New York City, Plenum, 1998

[3] Burke, N.E., Hobbs, R.E., Kashou, S.F., "Measurement and Modeling of Asphaltene Precipitation", *Journal of Petroleum Technology*, Volume 42, NO. 11, 1990

[4] Anderson, S.I. and Speight, J.G., "Thermodynamic Models for Asphaltene Solubility and Precipitation", *Journal of Petroleum Science and Engineering,* Volume 22, NO. 1-3, 1999

[5] Buenrostro-Gonzalez, E., Lira-Galeana, C., Gil-Villegas, A. et al., "Asphaltene Precipitation in Crude Oils: Theory and Experiments", *AIChE J.,* Volume 50, NO. 10, 2004

[6] Hoepfner, M.P., Limsakoune, V., Chuenmeechao, V. et al., "A Fundamental Study of Asphaltene Deposition", *Energy and Fuel*, Volume 27, NO. 2, 2013

[7] Gaspar, G. and Travalloni-Louvisse, A.M., "Adsorption of Asphaltenes and its Effect on Oil Production", *SPE Production and Facilities*, Volume 8, NO. 2, 1993

[8] Ali, M.F., Bukhari, A., Misbah-ul-Hasan, "Structural Characterization of Arabian Heavy Crude Oil Residue", *Fuel Science and Technology International*, Volume 7, NO. 8, 1989

[9] Fan, T., Buckley, J.S., "Rapid and Accurate SARA Analysis of Medium Gravity Crude Oils", *Energy Fuels*, Volume 16, NO. 6, 2002

[10] Peralta-Martínez, M.V., Vázquez-Ramírez, R., Blass-Amador, G. et al., "Determination of Functional Groups in Mexican Vacuum Residua", *Petroleum Science and Technology*, Volume 26, NO. 1, 2008

[11] Dolphin, D., "*The Porphyrins*", Vol. 1, Structure and Synthesis, Part A, New York, Academic Press, 1978

[12] Filby, R.H. and Branthaver, J.F., "*Metal Complexes in Fossil Fuels*", Symposium Series Number 344, American Chemical Society, Washington D.C. 1987

[13] Akmaz, S., Iscan, O., Gurkaynak, M.A. et al., "The Structural Characterization of Saturate, Aromatic, Resin, and Asphaltene Fractions of Batiraman Crude Oil", *Petroleum Science and Technology*, Volume 29, NO. 2, 2011

[14] Chamkalani, A., "Correlations between SARA Fractions, Density, and RI to Investigate the Stability of Asphaltene", *International Scholarly Research Network*, Volume 2012, Article ID 219276, 2011

[15] Prakoso, A.A., Punase, A.D., Hascakir, B., "A Mechanistic Understanding of Asphaltene Precipitation from Varying Saturate Concentration Perspective", Presented at SPE Latin American and Carribean Petroleum Engineering Conference, Quito, Ecuador, 2015

[16] Punase, A., Prakoso, A.A., Hascakir, B., "The Polarity of Crude Oil Fractions Affects the

Asphaltenes Stability", Presented at SPE Western Regional Meeting, Anchorage, Alaska, USA, 2016
[17] ASTM D2007-11, "Standard Test Method for Characteristic Groups in Rubber Extender and Processing Oils and Other Petroleum-Derived Oils by the Clay-Gel Absorption Chromatographic Method", West Conshohocken, PA: *ASTM International*, 2011
[18] Debye, P., "*Polar Molecules*", New York City, Dover, 1929
[19] Sidgwick, N.V., "Dipole Moment and Molecular Structure", *Chem. Rev.* Volume 19, NO. 3, 1936
[20] Prakoso, A.A., Klock, K., Punase, A. et al., "Determination of the Stability of Asphaltenes through Physicochemical Characterization of Asphaltenes", Presented at SPE Western Regional Meeting, Anchorage, Alaska, USA, 2016
[21] Bellamy, L.J., "*The Infrared Spectra of Complex Modules*", Vol. 2, London: Advances in Infrared Group Frequencies, Springer (Reprint), 1980
[22] Onsager, L., "Electric Moments of Molecules in Liquids", *J. Amer. Chem. Soc.*, Volume 58, NO. 8, 1936
[23] Smyth, J.W., "Some Developments of Guggenheim's Simplified Procedure for Computing Electric Dipole Moments", *Trans. Faraday Soc.* Volume 46, 1950
[24] Guggenheim, E.A., "The Computation of Electric Dipole Moments", *Trans. Faraday Soc.*, Volume 47, 1951
[25] Böttcher, C.J.F., "*Theory of Electric Polarization*", Amsterdam, Elsevier, 1952

7. Appendix

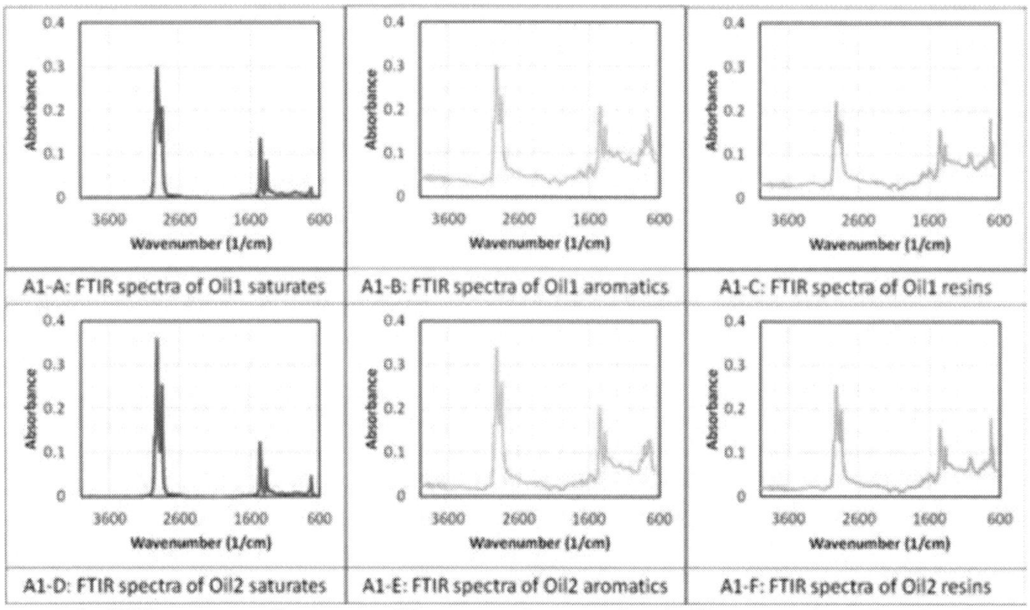

Figure A-1: FTIR spectra of saturates (blue curves), aromatics (yellow curves), and resins (green curves) of Oil1 and Oil2.

444380 Chemical Control of Asphaltene Aggregation from Cohesive Force Measurements

Abstract:

Asphaltene solids may precipitate in subsea crude oil wellbores and flowlines, due to the rapid decrease in fluid pressure during flow. While the consequence of asphaltene deposition and build-up at the wall is well-understood, little is known about how precipitated asphaltene particles behave in flow. In this work, a micromechanical force apparatus was used to estimate the asphaltene particle surface free energy in a variety of continuous phases; the apparatus was further able to capture the adsorption behaviour of surfactants at the asphaltene particle surface. Pentane-, hexane-, and heptane-extracted asphaltenes were tested in duplicate; one set of each was further washed with hot heptane to purify the extracted asphaltenes. The data suggests that asphaltene cohesive force does not exhibit a significant dependence on the extraction method of the asphaltenes, and indicates that asphaltene cohesion does not depend strongly on the carbon number of the continuous phase. In addition, we present the first data demonstrating adsorption-type behaviour of surfactants on the asphaltene particle surface, which demonstrates the effectiveness of chemical control of the asphaltene aggregation potential.

Copyright © American Institute of Chemical Engineers. All rights reserved.

444746 Effect of Emulsified Water on Asphaltene Instability in Crude Oils

Abstract:

Asphaltene precipitation and subsequent deposition during oil production are of great importance for the oil industry nowadays because of the potential risk associated to this heavy fraction in plugging wellbores and production equipment. Although water is commonly present in the produced fluids, because of instrument limitations and inadequate techniques, it is usually separated from the oil prior to any experimental analysis. Therefore, the effect of water on asphaltene stability and deposition tendency is not completely understood and the information available in the open literature is scarce. The focus of this work is on the determination of the effect of emulsified water on asphaltene instability in crude oil systems. Three crude oil samples, ranging from 26 to 40 °API, and 1.2 to 13 wt% asphaltene content, from Canada, Middle East and Gulf of Mexico, and one bitumen sample from the Athabasca region in Canada were used in this study. A total of nine systems were investigated with and without the presence of emulsified water. First the three crude oils were studied, and then model oil blends were prepared with extracted asphaltenes from the different oils using n-pentane as the asphaltene precipitant. Furthermore, three different asphaltene fractions were extracted from the Athabasca bitumen: the n-pentane insoluble (also known as n-C_5 asphaltenes), the n-heptane insoluble (or n-C_7 asphaltenes) and the n-pentane insoluble but n-heptane soluble fraction (n-$C_{5\text{-}7}$ asphaltenes). The stability of asphaltenes in the presence and absence of emulsified water was investigated by the indirect method. This is a multi-step method that includes the formation of an emulsion (when water is added), the precipitation of asphaltenes upon addition of a normal alkane for a specified aging period (24 hours in this case), centrifugation of the samples at 10,000 rpm for 25 min, and determination of the absorbance of the supernatant fluids at a wavelength of 1,100 nm. The centrifugation step is necessary to remove the water from the supernatant fluid and settle the asphaltene aggregates that are 100 nm and larger. The indirect method is used to determine the onset of asphaltene precipitation and how this is affected by the presence of emulsified water by analyzing the absorbance of the supernatant fluids as a function of the volume fraction of the normal alkane added to induce the precipitation. It was found that for the crude oils from the Middle East and Canada and their corresponding model oils, the addition of water neither did have a significant effect on the onset of asphaltene precipitation nor the amount of precipitated asphaltenes. However, the crude oil from Gulf of Mexico and the model oils from the Athabasca bitumen (containing n-C_5 and n-C_7 asphaltenes) were significantly affected by the presence of water. In the latter case, there is even a significant shift on the onset of asphaltene precipitation and the total amount of asphaltene precipitated. Interestingly, the model oil that was prepared with n-$C_{5\text{-}7}$ Athabasca asphaltenes did not show a significant change upon addition of water. The experimental evidence suggests that some asphaltenes are more prone to interact with water at the oil – water interface, which are in turn settled by the effect of centrifugation– than others. The difference between the affinity of asphaltenes from different sources for the oil – water interface and even within the polydisperse distribution of asphaltenes in a given oil sample is evident in this work. The effect of temperature on the water / asphaltene interactions in the crude oil from Gulf of Mexico was also investigated. It was found that at increasing temperatures, although the solubility of water in the oil phase does not change considerably, there is a significant effect on the drop of absorbance of the supernatant fluids after centrifugation, which we believe indicates an increased interaction between asphaltenes and water. With this study we aim to contribute to a better understanding of the interaction of water and asphaltenes in crude oil systems, which will eventually lead to the development of cost-effective strategies for the mitigation of this flow assurance problem.

Copyright © American Institute of Chemical Engineers. All rights reserved.

Mixture Effects on Adsorption Kinetics of Asphaltenes at the Water/Toluene Interface

Vincent Pauchard, Fang Liu

Energy Institute and Department of Chemical Engineering, The City College of New York, New York, NY
* Corresponding author e-mail: vpauchard@ccny.cuny.edu

Keywords: Adsorption Kinetics, Dilatational Rheology, Asphaltenes, Interface

ABSTRACT

Interfacial tension decay between water and organic phases containing asphaltenes is generally ascribed to slow interfacial reorganization/relaxation. One of the arguments allegedly supporting this conclusion is that the dynamics of this process seem way too slow compared to diffusion timescale as calculated from the bulk asphaltenes concentration. Such reasoning however neglects that asphaltenes are a solubility class comprising a wide variety of molecular structures, only a minority of them significantly adsorbing at the water/oil interface. As a consequence, the timescale of the global equilibration process is to be the diffusion timescale of the component present at the lowest concentration among those effectively populating the interface. To support this demonstration, experimental data of dynamical interfacial tension will be shown and prove to be accurately represented by using the same set of parameters extracted from its dilatational rheology based on a binary mixture model.

1. INTRODUCTION

Most of the researchers observed that the adsorption process of asphaltenes either to the oil-air interface or to the oil-liquid interface was not purely diffusion-controlled [1–4]. The previous pendant drop experiments using different concentrations of asphaltenes and different viscosities of solvents revealed that the adsorption of asphaltene at oil-water interface is mainly controlled by bulk diffusion at short time while the adsorption kinetics become slower and no longer diffusion-controlled at longer time [5]. They reported an asymptotic behavior of dynamic interfacial tension and surface coverage in long-term adsorption, which is much longer than the diffusion timescale [5].

Bauget et al. [3] conclude that the adsorption process involves a reorganization of asphaltenes molecules in a network structure, which further forms a solid film at the interface and stabilizes emulsion. This hypothesis is extensively supported by both shear and dilatational rheological studies [6–10]. However, this explanation that the deviation comes from conformational change of adsorbed asphaltene molecules on oil-water interface completely contradicts with the phenomena observed by Rane et al. [5]. At low concentrations (e.g. 10ppm), the surface coverage and surface pressure remains low. The curves of dynamic interfacial tension versus square root of time rescaled with viscosity superpose and match well with each other. While at high concentrations, the curve deviates from diffusion-controlled curve and the deviation appears earlier for solvent with lower viscosity [5,11], which should be in the opposite way in relaxation/reorganization theory.

The reasons for the discrepancy between diffusion-controlled adsorption kinetics and the measured long-term dynamic interfacial tension data still remain in debate. One possible explanation is that the asphaltene group is a class of polyaromatic hydrocarbons with diverse structure, which takes different mass fractions, various adsorption coefficients, and hence miscellaneous characteristic relaxation time [12–17]. The mixture of these asphaltene molecules may possess common natures such as being surfactant at oil-water interface and soluble in toluene but insoluble in *n*-heptane. Despite of those common natures, asphaltenes molecules are believed to have various adsorption coefficients and hence different relaxation timescales, which can be attributed to the difference in aromaticity, molecular weight, and polarity. The existence of heteroatoms also adds complexity to its physical behavior. Although several components have been characterized using high-resolution transmission electron microscopy (HRTEM) [18], high-resolution nuclear magnetic resonance (HRNMR) [19], X-ray photoelectron spectroscopy (XPS) [19,20], small angle neutron scattering (SANS) [21], some of the components are barely detected using current techniques. Binary mixture model using Ward-Tordai equation and Langmuir Equation of State (EOS) [22,23] has been developed earlier to account for the mixture effect. However, the adsorption curve exhibits bi-exponential decay and does not match monotonically with experimental data.

Hence, we proposed to extract a unique set of parameters from experimental data of dilatational rheology based on a binary mixture and then couple Langmuir EOS and diffusion kinetics to describe the experimental data of dynamic interfacial tension of the water-oil interface. This can be applied to any binary mixture of surfactants at liquid-liquid interface or air-liquid interface.

2. METHODOLOGY

2.1 The LVDT model

The behavior of the surface when the interface encounters a variation in area can be described by dilatational modulus:

$$E = A\frac{\partial \gamma}{\partial A} = -A\frac{\partial \Pi}{\partial A} \qquad \text{Equation 1}$$

where γ is surface tension, Π is surface pressure, and A is the interfacial area. Dilatational rheology of single surfactant/surfactant mixtures can be predicted from solving mass conservation of surfactant(s) in the system of bulk solution and adsorbed layer together with adsorption equilibrium between subsurface layer and adsorbed layer. Assuming diffusion-controlled adsorption mechanism (i.e. local equilibrium between the subsurface concentration of a surfactant and its adsorption), the analytical solution of dilatational modulus for a single surfactant at the subsurface layer has been derived by Lucassen and van den Tempel (LVDT) [24] as:

$$E = E_0 \times \frac{1+\zeta+i\zeta}{1+2\zeta+2\zeta^2} \qquad \text{Equation 2}$$

where the dimensionless number, ζ, is defined as $\zeta = \frac{1}{\sqrt{\omega\tau_D}}$ (where ω is the pulsation of oscillations and the characteristic diffusional relaxation time τ_D is $\tau_D = \frac{2}{D}(\frac{\partial \Gamma}{\partial c})^2$ with D and c the bulk diffusion coefficient and concentration, respectively) and the limiting elasticity E_0, is defined as $E_0 = \frac{d\Pi}{dln\Gamma}$. The original form of the solution can easily be separated into real/elastic part (E') and imaginary/viscous part (E'') with the introduction of the dimensionless number. Hence, limiting elasticity can be directly calculated from experimental moduli:

$$|E_0| = \frac{E'^2+E''^2}{E'-E''} \qquad \text{Equation 3}$$

Although the LVDT model was developed for equilibrium adsorption, this restriction can be negligible when the contribution of the ongoing macroscopic equilibration between bulk and interface is small compared to the change in surface coverage due to oscillations and local equilibration.

2.2 Binary mixture model

The analytical solution of dilatational modulus for binary mixture has been derived in previous studies [25–28]:

$$E = \frac{1}{B}\left\{\left(\frac{\partial\Pi}{\partial ln\Gamma_1}\right)_{\Gamma_2}\left[a_{11}\sqrt{\frac{i\omega}{D_1}} + a_{12}\frac{\Gamma_2}{\Gamma_1}\sqrt{\frac{i\omega}{D_2}} + (a_{11}a_{22} - a_{12}a_{21})\frac{i\omega}{\sqrt{D_1D_2}}\right] + \left(\frac{\partial\Pi}{\partial ln\Gamma_2}\right)_{\Gamma_1}\left[a_{22}\sqrt{\frac{i\omega}{D_2}} + a_{21}\frac{\Gamma_1}{\Gamma_2}\sqrt{\frac{i\omega}{D_1}} + (a_{11}a_{22} - a_{12}a_{21})\frac{i\omega}{\sqrt{D_1D_2}}\right]\right\} \quad \text{Equation 4}$$

where the coefficient $a_{ij} = \left.\frac{\partial\Gamma_i}{\partial c_j}\right|_{c_{k\neq j}}$ is a partial derivative which could be calculated from adsorption isotherm and the parameter B is given as $B = 1 + a_{11}\sqrt{\frac{i\omega}{D_1}} + a_{22}\sqrt{\frac{i\omega}{D_2}} + (a_{11}a_{22} - a_{12}a_{21})\frac{i\omega}{\sqrt{D_1D_2}}$. The limiting elasticity in this case becomes $E_0(\Gamma) = \Gamma\left.\frac{d\Pi}{d\Gamma}\right|_{\Gamma=\Gamma_1+\Gamma_2}$.

The adsorption isotherms available to account for mixture of asphaltenes at water-oil interface are based on ideal assumptions, as the distribution of different components is unknown. Langmuir adsorption isotherm is one among those proposed to describe adsorption of asphaltenes molecules at the oil-water interface. For a mixture of surfactants, the number of local adsorption sites per unit interfacial area, Γ_∞, is assumed to be same for all the components and the fractional surface coverage is defined as:

$$\theta_j = \frac{\Gamma_j}{\Gamma^\infty} = \frac{k_jc_j}{1+\sum k_jc_j} \quad \text{Equation 5}$$

where k_j is the adsorption coefficient of component j and c_j is the individual concentration of j. As can be observed from the equation, the product of each component's adsorption coefficient (k) and its concentration (c) should be comparable with each other so that the mixture effect is accounted for and each component could play a role. The adsorption coefficient of individual component can be then approximated to be inversely proportional to its concentration. Furthermore, it can be inferred that the majority of constituents in asphaltenes are not surface-active components, i.e. they have a low adsorption coefficient k.

The mixture model of binary system is proposed based on the above-mentioned correlation between adsorption coefficient and concentration. The binary model consists of one component with high surface activity (k_1) but low concentration (c_1) and the other one with low surface activity ($k_2 << k_1$) but high concentration ($c_2 >> c_1$).

Recent studies show that experimental data of surface pressure and elasticity of different types of asphaltenes at water-oil interface can be successfully described by Langmuir EOS [29,30]. Hence, a mixture Langmuir EOS [26] is selected to obtain the dynamic interfacial tension in short-time approximation:

$$\gamma(\Gamma) = \gamma_0 - kT\Gamma^\infty \ln\left(1 - \sum\theta_j\right) \quad \text{Equation 6}$$

where γ_0 is the interfacial tension of clean surface, θ_j is fractional surface coverage of surfactant j, and Γ^∞ for mixture surface excess coverage is defined as $\Gamma^\infty = \frac{\sum\theta_i}{\sum\theta_iA_i}$.

3. RESULTS AND DISCUSSION

3.1 LVDT model versus binary mixture model

To compare LVDT model for single surfactant system and binary mixture model (component 1 and 2), total concentration is taken as the same value, i.e. $c_{LVDT} = c_1 = c_2$. Diffusion coefficient of single surfactant and individual component in the binary mixture is assumed to be equal throughout the calculations ($D_{LVDT} = D_1 = D_2$), so is the excess surface coverage ($\Gamma_{LVDT}^\infty = \Gamma_1^\infty = \Gamma_2^\infty$). Under these assumptions, the adsorption coefficient in LVDT model is correlated to that of the individual component in the binary model as $k_{LVDT} = \frac{k_1c_1+k_2c_2}{c_1+c_2}$.

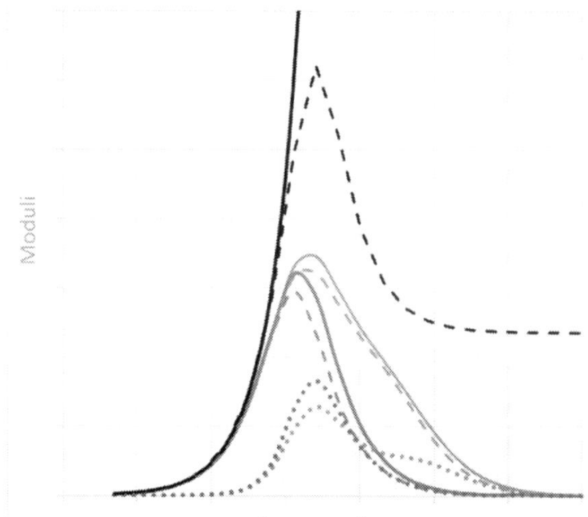

Figure 1: Complex moduli (solid lines), elastic moduli (dashed lines) and limiting elasticity (black curves; LVDT: solid; binary mixture model: dashed) as a function of concentration (LVDT model: blue; binary mixture model: red). Reprint from Liu et al. [31].

The dilatation moduli (complex, elastic and viscous) calculated from Equation 2 and 3 for a wide range of bulk concentrations is illustrated in Figure 1 above. For both LVDT model and binary mixture model, a bell-shaped curve is generated. At low concentration, complex modulus is low and mainly contributed by elastic modulus. Moreover, it is almost equal to limiting elasticity, which means it is nearly independent of frequency. As concentration increases, complex modulus increases and the same trend is observed in elastic and viscous modulus. The surface modulus remains frequency independent and close to limiting elasticity. Above a certain concentration (for which τ_D approaches $1/\omega$), modulus starts to decline and becomes frequency dependent. In this regime viscous modulus increases up to being equal to elastic one. As a consequence the phase shift between area and surface pressure oscillations ($\varphi = arctan\,(\frac{E''}{E'})$) increases from 0 at low concentration to 45 degrees at high concentration. Phase shift is also found to increase with concentration but always remain below 45 degrees regardless of concentration. These are the main features observed in both LVDT model and binary mixture model. Additionally, binary mixture model can be well fitted by an average LVDT model before the maximum modulus is reached while its deviation from the LVDT model starts to increase after the maximum point. This can be explained by the difference of the two surfactants in relaxation timescales, which is clearly observed when plotting moduli/phase shift as a function of frequency (see Figure 2).

As can be seen in Figure 2, moduli increases and phase shift decreases monotonically with increasing frequency in the LVDT model whereas in the binary mixture model, there are two cases to be discussed. When the relaxation timescales of two components do not differ much and the mixture can be considered as a continuum for molecular structure, a plateau appears in the evolution of phase shift curve at intermediate frequency and an inflection point occurs in the increase of modulus with frequency in the same region. These behaviors correspond to gel point rheology well where phase shift was also observed to be frequency dependent at both low and high frequency regime but frequency independent at intermediate regime. Moreover, the gel point region moves to higher frequency and higher phase shift values with higher concentration. On the other hand, if the timescales of two components vary by many orders of magnitude, phase shift reveals a bi-modal curve and modulus shows a stepwise increase with frequency. The phase angle first decreases with frequency and reaches a local minimum (it can be zero); then it increases up to a local maximum before decreasing again with frequency. In any case, phase angle is always lower than 45 degrees, which means the interface is more solid-like than liquid-like, and the phase angle is equal to $n \times 90$, where n is the order in the power law equation of critical gels. On the other hand, the complex modulus of surfactant mixture would always show an increase with frequency regardless of the composition.

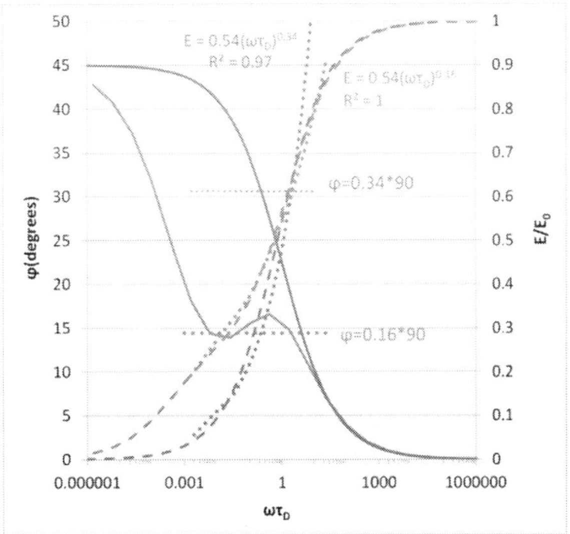

Figure 2: Moduli (dashed lines) and phase shift (solid lines) as a function of frequency (LVDT model: blue; binary mixture model: red). Both models are fitted using a power law equation (dotted lines) Reprint from Liu et al. [31].

3.2 Dilatational rheology and adsorption kinetics

For diffusion-limited adsorption, dynamic interfacial tension and dilatational moduli are governed by the same physics. Hence, it is possible to extract the properties of surfactant(s) from dilatational rheology experiments and use them to predict dynamic interfacial tension results. This is verified by applying the dilatational rheological experimental data from Sztukowski & Yarranton [32] on a concentration of 0.5 kg/m^3 asphaltenes in toluene, which is oscillated after 16 hours adsorption. To extract the mixture information from fitting the experimental data of moduli with the binary mixture model (Equations 4, 5, and 6), several parameters have to be known. The Langmuir EOS with a surface excess coverage Γ^∞ of 3.3 molecules/nm^2 has been retained from Rane et al. [29], which further imposes to use diffusion coefficient under the assumption of the infinite dilution ($D=2.5\times10^{-10}$ m^2/s was obtained by averaging values found in the literature [33,34]). Hence, the only unknowns are now subsurface concentrations after 16 hours' adsorption and adsorption coefficients of components 1 and 2 (c_{s1}, c_{s2}, k_1, and k_2) are there are eight experimental data points available to obtain these four variables. An additional constraint for optimization is the experimental surface pressure under the same conditions, namely, solvent, concentration, and time. Moreover, the optimized concentration should be located above the maximum point of modulus versus concentration curve so that each individual component can have an impact on rheology model instead of making a total effect. Under these conditions, the optimized variables are obtained as $c_{s1}=1.22\times10^{21}$ $molecules/m^3$, $k_1=$ 9.47×10^{-22} $m^3/molecule$, $c_{s2}=$ 3.08×10^{22} $molecules/m^3$, and $k_2=$ 2.13×10^{-23} $m^3/molecule$. The fitted values of modulus and phase shift together with experimental data are shown in Figure 3 below. As can be seen, the optimized model matches well with the experimental data points except for phase angle at $f=0.5$ Hz. The deviation of surface pressure is also considered negligible compared to the experimental value, which is not shown here. Those deviations are largely within experimental scattering. It can further be observed that both the experimental and theoretical data exhibit the gel point frequency dependence. It should also be kept in mind that the binary mixture model lumps all the various asphaltene molecules into two components and hence, the behaviors of those components under extreme conditions, i.e. the most and the least surface-active asphaltenes, are most probably not well captured.

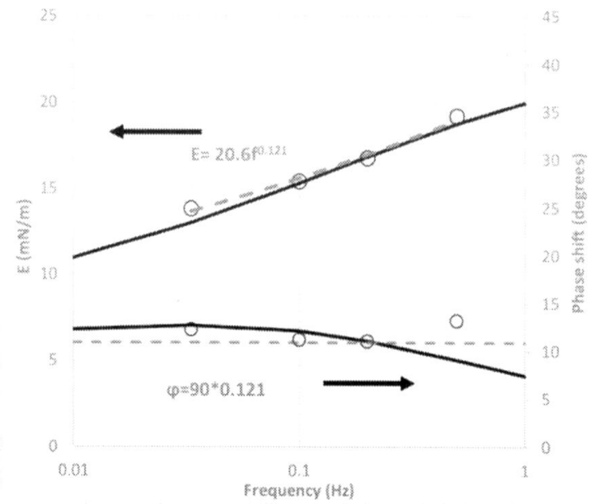

Figure 3: Fitting of experimental data from Sztukowski & Yarranton [32] (circles) by binary mixture model (solid lines) and power law equations (dotted lines). Reprint from Liu et al. [31]

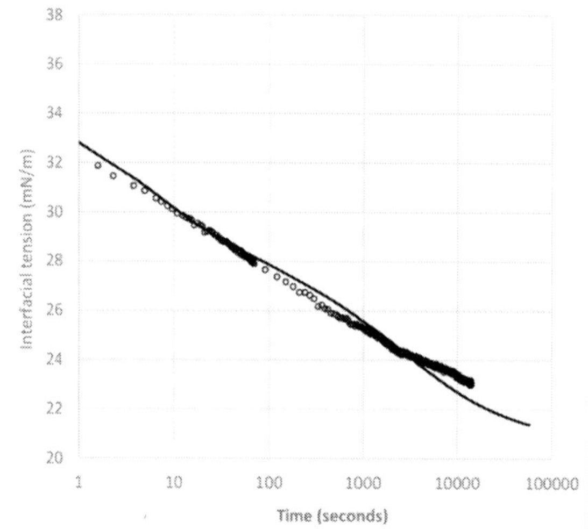

Figure 4: Comparison of predicted (line) and experimental (circles, data from Sztukowski & Yarranton [32]) dynamic interfacial tension for 0.5 kg/m^3 asphaltenes in toluene. Clean surface interfacial tension was taken as 35 mN/m [32]. Reprint from Liu et al. [31]

In order to compare with experimental data of dynamic interfacial tension, individual bulk concentrations c_1 and c_2 have to be determined from their optimized subsurface concentrations after 16 hours' adsorption. To start with, the optimized c_{s1} and c_{s2} values can serve as a first guess for c_1 and c_2 and then Ward and Tordai equation [35] is employed to calculate dynamic surface coverage. For component 2 who has a higher concentration but lower surface activity, its subsurface concentration reaches 99% of bulk concentration within 700 seconds, which validates the approximation $c_2 \approxeq c_{s2}$. With respect to component 1 with a much lower concentration, its subsurface concentration only reaches 75% of bulk concentration after 16 hours, which means the value of c_1 has to be progressively increasing in order to reach the subsurface concentration c_{s1} after 16 hours' adsorption. With these conditions and surface excess coverage value of 3.3 molecules/nm^2, dynamic interfacial tensions are calculated using Equation 6 and is shown in Figure 4 above with the experimental data provided by Sztukowski & Yarranton [32]. As can be seen, the predicted curve fits well with the experimental data. Meanwhile, the binary mixture model exhibits some inflection points that are not present in the experimental data, which may originate from the fact that asphaltenes with all their diversity are lumped into only two components. This can also explain the deviation between the binary mixture model and the experimental data at the initial stage of adsorption process where the effect of less surface-active surfactants is neglected in the binary model.

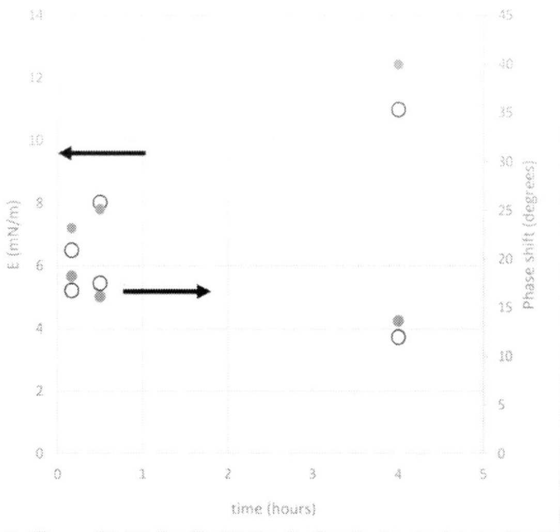

Figure 5: Time evolution of experimental dilatational rheology (open circles, 0.5 kg/m^3 asphaltenes in toluene, frequency of 0.033 Hz, data from Sztukowski & Yarranton [32]) and predicted dilatational rheology from Equations 4 to 6 (closed circles). Reprint from Liu et al. [31]

Furthermore, the adsorption kinetics can be used to predict the evolution of interfacial rheology with time under the same conditions (0.5 kg/m^3 asphaltenes in toluene). As shown in Figure 5, for 10 minutes, 30 minutes and 4 hours, the predicted modulus is within 1.5 mN/m of experimental value and the predicted phase angle is within 2 degrees of experimental one, which prove to be good predictions of experimental data.

The same analysis can be performed for other concentrations present in the same paper [32] and yield similar results. However, the optimized values of k_1 and k_2 are not strictly the same for all concentrations and the composition of the binary mixture, i.e. ratio of c_1 over c_2, is also varying, which is not desired. There are several reasons. When the bulk concentration is lowered, the influence of the asphaltenes with low surface activity over the experimental time range will be more dominant. Moreover, the influence of nanoaggregation on the optimized subsurface concentration and surface activity of available asphaltene monomers might be very complex. In particular, if the most surface-active asphaltenes are those with the aromatic core having the strongest electron donor/acceptor behavior, they might as well be those the most prone to aggregate.

4. CONCLUSIONS

Experimental data from dilatational rheology and adsorption kinetics of surfactant mixture are correlated by implementing a unique set of parameters, i.e. subsurface concentration and adsorption coefficient of individual components. These parameters obtained from fitting the experimental data by a binary mixture model in dilatational rheology can be coupled with selected adsorption isotherms and equation of state to predict the dynamic interfacial tension data. This method proves to be effective and the key features identified from data analysis of LVDT model and binary mixture model simulation are verified in the published data.

5. REFERENCES

1. Sheu, E., Storm, D. A. & Shields, M. B. Adsorption kinetics of asphaltenes at toluene/acid solution interface. *Fuel* **74,** 1475–1479 (1995).
2. Sheu, E. Y., De Tar, M. M. & Storm, D. A. Interfacial properties of asphaltenes. *Fuel* **71,** 1277–1281 (1992).
3. Bauget, F., Langevin, D. & Lenormand, R. Dynamic Surface Properties of Asphaltenes and Resins at the Oil–Air Interface. *J. Colloid Interface Sci.* **239,** 501–508 (2001).
4. Jeribi, M., Almir-Assad, B., Langevin, D., Hénaut, I. & Argillier, J. F. Adsorption Kinetics of Asphaltenes at Liquid Interfaces. *J. Colloid Interface Sci.* **256,** 268–272 (2002).
5. Rane, J. P., Harbottle, D., Pauchard, V., Couzis, A. & Banerjee, S. Adsorption Kinetics of Asphaltenes at the Oil–Water Interface and Nanoaggregation in the Bulk. *Langmuir* **28,** 9986–9995 (2012).
6. Spiecker, P. M. & Kilpatrick, P. K. Interfacial Rheology of Petroleum Asphaltenes at the Oil−Water Interface. *Langmuir* **20,** 4022–4032 (2004).
7. Harbottle, D. *et al.* Problematic stabilizing films in petroleum emulsions: Shear rheological response of viscoelastic asphaltene films and the effect on drop coalescence. *Langmuir* **30,** 6730–6738 (2014).
8. Verruto, V. J., Le, R. K. & Kilpatrick, P. K. Adsorption and molecular rearrangement of amphoteric species at oil-water interfaces. *J. Phys. Chem. B* **113,** 13788–13799 (2009).
9. Bouriat, P., El Kerri, N., Graciaa, A. & Lachaise, J. Properties of a Two-Dimensional Asphaltene Network at the Water−Cyclohexane Interface Deduced from Dynamic Tensiometry. *Langmuir* **20,** 7459–7464 (2004).
10. Dicharry, C., Arla, D., Sinquin, A., Graciaa, A. & Bouriat, P. Stability of water/crude oil emulsions based on interfacial dilatational rheology. *J. Colloid Interface Sci.* **297,** 785–791 (2006).
11. Pauchard, V., Rane, J. P., Zarkar, S., Couzis, A. & Banerjee, S. Long-Term Adsorption Kinetics of Asphaltenes at the Oil–Water Interface: A Random Sequential Adsorption Perspective. *Langmuir* **30,** 8381–8390 (2014).
12. Acevedo, S., Borges, B., Quintero, F., Piscitelly, V. & Gutierrez, L. B. Asphaltenes and Other Natural Surfactants from Cerro Negro Crude Oil. Stepwise Adsorption at the Water/Toluene Interface: Film Formation and Hydrophobic Effects. *Energy & Fuels* **19,** 1948–1953 (2005).
13. Fossen, M., Kallevik, H., Knudsen, K. D. & Sjöblom, J. Asphaltenes Precipitated by a Two-Step Precipitation Procedure. 2. Physical and Chemical Characteristics. *Energy & Fuels* **25,** 3552–3567 (2011).
14. Yang, F. *et al.* Asphaltene Subfractions Responsible for Stabilizing Water-in-Crude Oil Emulsions. Part 1: Interfacial Behaviors. *Energy & Fuels* **28,** 6897–6904 (2014).
15. Yarranton, H. W., Hussein, H. & Masliyah, J. H. Water-in-Hydrocarbon Emulsions Stabilized by Asphaltenes at Low Concentrations. *J. Colloid Interface Sci.* **228,** 52–63 (2000).
16. Schuler, B., Meyer, G., Peña, D., Mullins, O. C. & Gross, L. Unraveling the Molecular Structures of Asphaltenes by Atomic Force Microscopy. *J. Am. Chem. Soc.* **137,** 9870–9876 (2015).
17. Speight, J. G. Petroleum Asphaltenes - Part 1: Asphaltenes, Resins and the Structure of Petroleum. *Oil Gas Sci. Technol.* **59,** 467–477 (2004).

18. Sharma, A., Groenzin, H., Tomita, A. & Mullins, O. C. Probing Order in Asphaltenes and Aromatic Ring Systems by HRTEM. *Energy & Fuels* **16,** 490–496 (2002).
19. Andrews, A. B. *et al.* Comparison of coal-derived and petroleum asphaltenes by 13C nuclear magnetic resonance, DEPT, and XRS. *Energy and Fuels* **25,** 3068–3076 (2011).
20. Rudrake, A., Karan, K. & Horton, J. H. A combined QCM and XPS investigation of asphaltene adsorption on metal surfaces. *J. Colloid Interface Sci.* **332,** 22–31 (2009).
21. Spiecker, P. M., Gawrys, K. L., Trail, C. B. & Kilpatrick, P. K. Effects of petroleum resins on asphaltene aggregation and water-in-oil emulsion formation. *Colloids Surfaces A Physicochem. Eng. Asp.* **220,** 9–27 (2003).
22. Van den Bogaert, R. & Joos, P. Diffusion-controlled adsorption kinetics for a mixture of surface active agents at the solution-air interface. *J. Phys. Chem.* **84,** 190–194 (1980).
23. Miller, R., Fainerman, V. B. & Aksenenko, E. V. A simple method to estimate the dynamic surface pressure of surfactant mixtures. *Colloids Surfaces A Physicochem. Eng. Asp.* **242,** 123–128 (2004).
24. Lucassen, J. & Van Den Tempel, M. Dynamic measurements of dilational properties of a liquid interface. *Chem. Eng. Sci.* **27,** 1283–1291 (1972).
25. Aksenenko, E. V., Kovalchuk, V. I., Fainerman, V. B. & Miller, R. Surface dilational rheology of mixed surfactants layers at liquid interfaces. *J. Phys. Chem. C* **111,** 14713–14719 (2007).
26. Fainerman, V. B. *et al.* Adsorption layer characteristics of Tritons surfactants. 3. Dilational visco-elasticity. *Colloids Surfaces A Physicochem. Eng. Asp.* **334,** 16–21 (2009).
27. Fainerman, V. B. *et al.* Adsorption Layer Characteristics of Mixed SDS/C_nEO_m Solutions. II. Dilational Viscoelasticity. *Langmuir* **26,** 1796–1801 (2010).
28. Alahverdjieva, V. S. *et al.* Competitive Adsorption from Mixed Hen Egg-White Lysozyme/Surfactant Solutions at the Air–Water Interface Studied by Tensiometry, Ellipsometry, and Surface Dilational Rheology. *J. Phys. Chem. B* **112,** 2136–2143 (2008).
29. Rane, J. P., Pauchard, V., Couzis, A. & Banerjee, S. Interfacial rheology of asphaltenes at oil-water interfaces and interpretation of the equation of state. *Langmuir* **29,** 4750–4759 (2013).
30. Rane, J. P. *et al.* Applicability of the Langmuir Equation of State for asphaltene adsorption at the oil-water interface: coal-derived, petroleum and synthetic asphaltenes. *Energy & Fuels* **29,** 3584–3590 (2015).
31. Liu, F., Darjani, S., Banerjee, S., Maldarelli, C. & Pauchard, V. Mixture effect on dilatation rheology of asphaltenes laden interfaces. *Langmuir*, manuscript submitted.
32. Sztukowski, D. M. & Yarranton, H. W. Rheology of Asphaltene–Toluene/Water Interfaces. *Langmuir* **21,** 11651–11658 (2005).
33. Lisitza, N. V, Freed, D. E., Sen, P. N. & Song, Y.-Q. Study of Asphaltene Nanoaggregation by Nuclear Magnetic Resonance (NMR). *Energy & Fuels* **23,** 1189–1193 (2009).
34. Norinaga, K., Wargardalam, V. J., Takasugi, S., Iino, M. & Matsukawa, S. Measurement of Self-Diffusion Coefficient of Asphaltene in Pyridine by Pulsed Field Gradient Spin–Echo ^1H NMR. *Energy & Fuels* **15,** 1317–1318 (2001).
35. Ward, A. F. H. & Tordai, L. Time-Dependence of Boundary Tensions of Solutions I. The Role of Diffusion in Time-Effects. *J. Chem. Phys.* **14,** 453 (1946).

445325 A New Asphaltene Test Under High Temperature and High Pressure – Asphaltene Rocking Cell

Abstract:

The most widely used asphaltene tests in the laboratory are based on solubility and precipitation. Those methods such as Turbiscan and Asphaltene Dispersant Test (ADT) have limitations in operating at high temperature, high pressure and under water incorporated conditions. These limitations might lead to unrealistic interpretation of asphaltene behavior and optimal chemical selections.

Nalco Champion has developed a new method of asphaltene testing which accommodates high temperature and high pressure conditions. This Asphaltene Rocking Cell (ARC) test is analogous of the well-known Hydrate Rocking Cell test but able to operate in high temperature and high pressure (HT/HP) conditions, up to 15,000psi and 250°F. ARC provides pseudo-live oil condition by combining synthetic field gas with dead crude oil in HT/HP conditions then asphaltene destabilization is demonstrated via depressurization. This enables understanding for asphaltene behavior and the assessment of chemical performances under field conditions. ARC test can be operated in various conditions not only in HT/HP but also in various Gas to Oil Ratios (GOR), different water-cuts and brine compositions. ARC also has advantage on asphaltene testing in CO_2 flood field sample. Nalco Champion has successfully utilized ARC in several case studies from Gulf of Mexico assets and on-shore CO_2 flood assets to deliver best chemical solutions for asphaltene problems which were difficult to achieve by conventional asphaltene testing methods.

In this paper, we will discuss about the ARC test method development and provide case histories of successes of utilizing ARC.

Copyright © American Institute of Chemical Engineers. All rights reserved.

444833 On the Evaluation of the Performance of Asphaltene Dispersants

Abstract:

Asphaltene precipitation and subsequent deposition is a potential flow assurance problem for the oil industry nowadays. Moreover, because oil production is moving to more difficult production environments – e.g. deeper waters – or is focusing on extracting residual oil using enhanced oil recovery techniques, the significant changes of pressure, temperature and/or composition can aggravate the asphaltene deposition problems. One of the most common strategies to prevent or at least reduce asphaltene deposition is the utilization of chemical additives. However, there are still several unresolved challenges associated to the utilization of these chemicals: First, the experimental conditions and results obtained in the lab are not always consistent with the field observations. Also, in some cases these chemical additives seem to worsen the deposition problem in the field. Therefore, there is a clear need to revisit the commercial techniques that are used to test the performance of asphaltene inhibitors and to provide a better interpretation of the results obtained. In this work, a technique based on NIR spectroscopy is presented to evaluate the performance of three commercial asphaltene dispersants. The results are also validated using digital optical microscopy. The method presented in this work is faster and more reproducible compared to the available methods such as the Asphaltene Dispersion Test (ADT) and the Solid Detection System (SDS). Also, unlike the ADT test, our proposed method can evaluate the performance of the dispersants in a wide range of temperatures and compositions. The experimental evidence shows that the asphaltene dispersants neither shift the actual onset of asphaltene precipitation nor reduce the amount of asphaltene precipitated. We believe that some results that have been reported that suggest that asphaltene dispersants can actually shift the onset of asphaltene precipitation are an unfortunate combination of insufficient sensitivity of the commercial instruments used and the slowing down of the asphaltene aggregation process by the effect of the added dispersants. The chemical additive dosage, aging time and temperature effect on the asphaltene aggregation process are also discussed in this manuscript.

With this work we aim to contribute to a better understanding of the variables that affect the performance of asphaltene dispersants, and the effect that these chemicals have on the complex multi-step mechanism of asphaltene precipitation, aggregation and deposition.

Copyright © American Institute of Chemical Engineers. All rights reserved.

443325 An Industrial Vision of Asphaltene Risk during Oil Production

Abstract:

Efficient development strategies from reservoir to topside process chain require a perfect knowledge of the phase behavior of the produced fluids. Among these fluids, the ones containing asphaltene have a very singular position.

Asphaltenes have been studied for decade, as their very peculiar properties open a quasi infinite number of interesting studies to be performed, either from an academic perspective or from a more practical and operational point of view.

The most conventional way to describe the behavior of asphaltenes consider that, at reservoir conditions at high pressure, asphaltenes are stabilized in solution through the high density of the oil. As the pressure decreases during production, the density can decrease ultimatly to the point where the solvency power of the fluid phase is not good enough, so that asphaltenes start to flocculate. It is usually considered that this flocculation onset pressure has to be at least above the saturation pressure, as below that pressure the gas / liquid phase separation will cause the density of the liquid phase to increase again. It is particularly true if the nature of the gas does not influence the stability of asphaltenes by other mechanisms than its role of the density of the oil. Asphaltene risk is thus a complex interplay between the solvency power of the oil and the nature of the asphaltenes, and between thermodynamic and kinetics effects.

Ideally, the perfect workflow for flow assurance studies linked to asphaltene deposition would require 3 steps :

1/ to be able to predict the risk from a thermodynamic standpoint,

2/ to be able to quantify this risk through the integration of the kinetics effects, and

3/ to determine how to efficiently mitigate this risk, in particular through the use of chemical additives.

Recent developments or projects tend to raise the difficulty of achieving efficient predictions, by complexifing the systems, through higher temperature, pressure and nature of the gases, and commingling of different fluids.

By comparing our internal industrial practices to what can be found in the open academic literature, we will share our view of what is well understood, and the points that are still to be improved.

Copyright © American Institute of Chemical Engineers. All rights reserved.

443236 Rethinking the Modeling Approach for Asphaltene Precipitation Using the PC-SAFT Equation of State

Abstract:

The Perturbed Chain version of the Statistical Associating Fluid Theory Equation of State (PC-SAFT EoS) has been successfully applied to the modeling of the phase behavior of petroleum systems at reservoir conditions and the prediction of asphaltene precipitation. Although the predictive capabilities of this method and its wide range of applications are much better than other modeling techniques based on conventional cubic equations of state, there are certain features that are still missing, such as the polydisperse nature of asphaltenes. Furthermore, the simulation parameters are usually tuned to Asphaltene Onset Pressure (AOP) data –which are usually obtained by depressurizing live oil samples– or titration experiments of dead oil samples at ambient conditions. These techniques are subject to significant uncertainties caused by slow kinetics of asphaltene aggregation and the limited sensitivity of the instruments used. The focus of this work is on the experimental study and modeling of asphaltene precipitation from n-alkane - diluted crude oil mixtures considering the polydispersity of asphaltenes. To investigate the effect of asphaltene polydispersity, a crude oil sample is studied by sequential precipitation with different normal alkanes. A recently developed experimental technique called the "Indirect Method" is used for the detection and quantification of asphaltene precipitation. Based on our previous research using the indirect method, it was found that both the onset and the amount of asphaltene precipitation change with the aging time for a wide range of volume fractions of the added n-alkane. However, the data showed that the amount of precipitated asphaltene at 90 volume percent of a given precipitant is nearly time invariant. This experimental data point is not controlled by the kinetics of asphaltene aggregation and, for this reason, is used in this study to adjust the simulation parameters. The experimental data are modeled using the Perturbed Chain form of the Statistical Associating Fluid Theory. An enhanced characterization method is used to minimize the number of adjustable parameters. In this work, a distribution function is proposed to represent the asphaltene molecular weight distribution for different asphaltene sub-fractions. It has been found that the inclusion of four asphaltene sub-fractions to represent the polydispersity of asphaltenes is necessary to match the amount of asphaltene precipitated. The modeling results show a reasonable agreement for the precipitated amount at 90 volume percent of n-pentane, n-hexane, n-heptane and n-octane. The PC-SAFT EoS predicts an onset point that requires less addition of n-alkane than the corresponding experimental point after one day of aging time. This result suggests that a thermodynamic limit exists for the onset of asphaltene precipitation and that it corresponds to experiments that require very long aging times. By fitting the simulation parameters to the amount of precipitated asphaltenes (and not the detected onset points) at 90 volume percent of added asphaltene precipitant, we can successfully remove the variability of the data with respect to the aging time. This tuning procedure also enables a more accurate prediction of the amount of precipitated asphaltenes at reservoir conditions, which is a more important result for the modeling of asphaltene deposition than the Asphaltene Onset Point itself. The integration of this novel characterization procedure as part of the Asphaltene Deposition Tool (ADEPT) developed at Rice University is a topic of ongoing research.

Copyright © American Institute of Chemical Engineers. All rights reserved.

AUTHOR INDEX

AUTHOR INDEX

AIChE
120 Wall Street, FL 23
New York, NY 10005-4020

ISBN 978-1-5108-2489-8

9 781510 824898

CURRAN ASSOCIATES INC.
proceedings
.com